Goncalves Albino Dauala
João Raimundo Chigarisso
Regina Gonçalves Aleixo

EMPIRICAL METHODS FOR ESTIMATING REFERENCE EVAPOTRANSPIRATION

Goncalves Albino Dauala
João Raimundo Chigarisso
Regina Gonçalves Aleixo

EMPIRICAL METHODS FOR ESTIMATING REFERENCE EVAPOTRANSPIRATION

IRRIGATION MANAGEMENT

ScienciaScripts

Imprint

Any brand names and product names mentioned in this book are subject to trademark, brand or patent protection and are trademarks or registered trademarks of their respective holders. The use of brand names, product names, common names, trade names, product descriptions etc. even without a particular marking in this work is in no way to be construed to mean that such names may be regarded as unrestricted in respect of trademark and brand protection legislation and could thus be used by anyone.

Cover image: www.ingimage.com

This book is a translation from the original published under ISBN 978-620-6-75906-5.

Publisher:
Sciencia Scripts
is a trademark of
Dodo Books Indian Ocean Ltd. and OmniScriptum S.R.L publishing group

120 High Road, East Finchley, London, N2 9ED, United Kingdom
Str. Armeneasca 28/1, office 1, Chisinau MD-2012, Republic of Moldova, Europe
Printed at: see last page
ISBN: 978-620-7-86604-5

DECLARATION OF HONOR

I, **João Raimundo Chigarisso,** declare that this monograph is the result of my own work and is being submitted for the degree of Bachelor of Environmental Agricultural Engineering at Zambezi University, Chimoio. It has not been submitted before for any degree or for assessment at any other university.

____

(João Raimundo Chigarisso)

Chimoio, ______ November 2020

DEDICATORY

I dedicate it to my parents Raimundo Chigarisso and Luísa João.

When everything you try seems to be going wrong, remember that the plane takes off against the wind, not in its direction.

"Hery Ford"

ACKNOWLEDGMENTS

First of all, I would like to thank God for his infinite grace and mercy in my life and for always being by my side and encouraging me, granting me health, wisdom, deliverance and wonderful victories, to him and for him all things were created, so to him be all the glory forever.

I would like to thank my parents, Raimundo Chigarisso and Luísa João, for bringing me into this world. I owe you everything I've achieved in all this time. To my siblings Manuel Raimundo Chigarisso, Domingas Raimundo Chigarisso and Madalena Raimundo Chigarisso, for their support, strength, courage and for believing in me, I would like to say a big thank you. To my aunt and uncle João Sebastião Chigarisso, my uncle Joaquim Paia and my aunt Arminda Chigarisso for always giving me strength and advice, to my brother-in-law Simone Satiel and all the family in general I thank you for your strength and support.

I would like to thank the FEARN professors who have contributed their technical and scientific knowledge directly or indirectly.

To the supervisor and co-supervisor, Eng. Gonçalves Albino Dauala (MSc) and Dr. Jemusse Nobosse Noé Tesse Jemusse, who guided this monograph from the beginning, I would like to thank you for your patience, dedication, availability and rich advice, which certainly helped me grow. It was an honor to be supervised by you and to work alongside such competent professionals was a great pleasure.

I would like to thank my colleagues in Environmental Agricultural Engineering and friends, Anselmo Patricio, Assis Viramão, Halide Macuede, Gloria Marile, Enia Caanai, Fgrancisco Jeito, Honisio Manuel, Lingson Virgílio, Yardy Covane, José Baptista and Samuel Mandala for sharing science. Your friendship is invaluable, I will always carry all the memories with me.

CONTENTS

SUMMARY

Determining reference evapotranspiration (ETo) is essential for proper water management in agriculture, as it is the main parameter used to estimate crop water requirements for proper irrigation management. The aim of this study was therefore to compare thirty empirical methods for estimating reference evapotranspiration on a daily scale in relation to ETo obtained by the Piche evaporimeter, estimated by the Hargreave-Samani method (1985), Penman-Monteith adjusted by Oliveira and Carvalho (1998), in order to find the simplest method for estimating ETo and one that is accurate for irrigation management. A series of meteorological data obtained from the National Institute of Meteorology (INAM) Manica Province Delegation from October 2019 to March 2020 was used. To identify the best empirical method for estimating ETo, the agreement index (d) of Willmot et al. (1985), correlation coefficient (r), coefficient of determination (R^2), CS index of Camargo and Sentelhas (1997), root mean square error (RQME) and mean absolute error (MAS) were evaluated. The different methods analyzed showed low precision, low correlation and poor performance when compared to the ETo obtained by the Piche method. While the methods of Hargreave modified by Bert *et al.,* (2014), Linacre (1977), Benevides Lopes (1970), Hamon (1961), and Hargreave modified showed good performance when compared to the method of Hargreave-Samani (1985) and Penman-Monteith adjusted by Oliveira and Carvalho (1998). Therefore, for irrigation management we recommend estimating ETo using the methods that performed well in this study.

Keywords: empirical methods, reference evapotranspiration and irrigation management.

ABSTRACT

The determination of evapotranspiration reference (ETo) is essential for Nature's Way the run c to water in agriculture, it is the main parameter used to estimate needs water crops for proper management of irrigation. Thus, the objective of this study was to compare thirty empirical methods of estimating reference evapotranspiration on the daily scale in relation to the ETo obtained by the Piche aporimeter , estimated by the Hargreave-Samani method (1985), PM adjusted by Oliveira and Carvalho (1998) , in order to find the simplest method for estimating ETo and presenting good precision in irrigation management. He used a series of meteorological data obtained from the National Institute of Meteorology (INAM) Delegation of Manica Province October 2019 to March 2020. To identify the best Meto of ETo estimation empirical , evaluated the concordance index (d) by Willmot et al. (1985) , correlation coefficient (r) , determination coefficient (R^2), Camargo and Sentelhas CS index (1997) , mean square root of the error (RQME) and mean absolute error (EAM).The different methods analyzed showed low precision, correlation and poor performance when compared with ETo obtained by the standard method of Tar. While the methods of Hargreave modified by Bert *et al.* (2014) , Linacre (1977), Benavides Lopes (1970) and Hamon (1961) and Hargreave modified presented good performance when compared with Hargreave-Samani (1985) and PM set by Oliveira and Carvalho (1998) . Therefore, for irrigation management it is recommended to estimate ETo using the methods that performed well in the present study.

Key-words: empirical methods, reference evapotranspiration and irrigation management

LIST OF ACRONYMS, SYMBOLS AND ABBREVIATIONS

INAM: National Institute of Meteorology

ETo: Reference evapotranspiration

ETc: Crop evapotranspiration

Kc: Crop coefficient

FAO: United Nations Food and Agriculture Fund

WMO: World Meteorological Organization

CS: Camargo and Sentelhas Coefficient (1997)

MAS: Mean Absolute Error

P-M: Penman-Monteith

RQME: Root Mean Square Error

R^2 : Coefficient of determination

Ro: solar radiation incident at the top of the atomosphere

p: monthly percentage of electricity

ea: vapor pressure

es: saturation vapor pressure

Td: dew point temperature

h: altitude

lat: latitude

z: installation height of the wind speed measuring instrument

u_2 : average air speed measured at a height of 2 meters from the surface

r: Pearson's linear correlation coefficient

d: Wilmott *et al.* concordance index (1985)

Yi = ETo values estimated by the standard method

Xi = ETo values estimated by the other methods

X = average of the ETo values estimated by the other methods

Y = average of the ETo values estimated by the standard method

n = number of days in the period

I. INTRODUCTION

1.1. Background

Agricultural activity is highly dependent on the climatic conditions in the various production regions and is significantly influenced by various meteorological factors such as rainfall, solar radiation and wind speed (SILVA *et al.* 2014). For Borges and Mendiondo (2007) it is essential to study the influence of variations on the different strategies for using the agricultural system, in order to provide subsidies for the decision-making process of how much and when to water. Logchem and Queface (2012) state that one of the main factors restricting crop yields in Mozambique is crop evapotranspiration (ETc).

Evapotranspiration is quantified using the water balance of the layer occupied by the plant's root system, with evapotranspiration and precipitation being its main components (MENDONÇA *et al.*, 2003). A simpler alternative for determining crop evapotranspiration (ETc) is through the procedure suggested by the FAO (ALLEN *et al.*, 1998), in which the reference evapotranspiration (ETo) is multiplied by the crop coefficient (Kc).

ETo was first defined by Doorenbos and Pruitt (1977) as the evapotranspiration of an extensive area covered by actively growing grass covering the entire soil, with a height of 0.08 m to 0.15 m, without water restriction, free of disease and with an adequate border (ALLEN et al., 1998). In 1990, experts in agrometeorology proposed another concept of ETo, as being that observed from a hypothetical crop with a height of 0.12 m, aerodynamic resistance of the surface of 70 sm^{-1} and albedo of 0.23. This parameterization was included in the Penman-Monteith equation and was considered the standard (ALLEN et al., 1998 and ARAÚJO *et al.*, 2007).

Estimating ETo is important for identifying temporal variations in irrigation demand and for calculating water requirements for irrigation (BERNARDO *et al.*, 2013 and MENDONÇA, 2003). According to Carvalho and Oliveira (2012), quantifying ETo is also important in non-irrigated agriculture, as it allows sowing times to be planned according to the average water availability of the region in question, enabling greater efficiency in the use of rainfall.

Bernado *et al.* (1996) report that ETo can be determined by direct and indirect methods, with direct methods using lysimeters, experimental plots in the field, soil moisture control and the method of water inflow and outflow over large areas, with the lysimeter being the most accurate. Indirect methods estimate evapotranspiration using mathematical models that require knowledge of meteorological variables (MATOS and SILVA 2016). These models can be simple (based on temperature), such as Thornthwaite's in 1948 (PEREIRA *et al.*, 1997), or more complex, involving the energy balance, such as Penman-Monteith (ALLEN *et al.*, 1998).

The method considered the standard for estimating ETo is the Penman-Monteith method (FAO-56), which performs well in different types of climate and its value is very close to the ETo of the grass in the different types of climate tested. It has a physical basis and incorporates both physiological and aerodynamic parameters (PEREIRA *et al.*, 1997; CARLESSO *et al.*, 2007). The difficulty in using this method is that it requires measurements of many meteorological variables, such as air temperature, relative humidity, solar radiation balance, ground heat flux and wind speed, which are often not available in any property, region or country (CARLESSO *et al.*, 2007; CARVALHO and OLIVEIRA, 2012). According to Camargo and Camargo, (2000) when all this meteorological data is not available, an alternative is to use simpler empirical methods with fewer input variables.

Of the methods that use only air temperature as input data, the Hargreaves and Samani (1985) method performed best when compared to the standard Penman-Monteith FAO-56 method in the Aquidauana-MS region (JÚNIOR *et al.*, 2012). This same method performed very well in the region of Sobral, CE, which can be explained by the variation in temperature and solar radiation in the area (GONÇALVES et al., 2009). For the municipality of Paraipaba, CE, this method was classified as Poor, a controversial result compared to other authors, which can be explained by the low variation in temperature and solar radiation in that region (MEDEIROS, 2002). It tends to overestimate the ETo value, especially in humid climates, requiring regional calibration to adjust its accuracy (MANTOVANI et al., 2013). Camargo and Sentelhas (1997) found that on a monthly scale, this method overestimates the reference evapotranspiration measured in drainage lysimeters cultivated with grass for the state of São Paulo. Alencar et al. (2015) found that on a monthly scale this method performed worse for the climatic conditions of the municipality of Uberaba in estimating ETo when compared to the PM-FAO method.

Another method that can be used in a simpler way is the Class A Tank method. According to Mantovani et al. (2013) this method has technical limitations, especially for high frequency (localized) irrigation, and the results are usually less accurate than methods based on air temperatures and are more subject to external problems such as animals and leaks. For Stone and Silveira, (1995) although the method is practical and economical, it is less accurate due to the lack of similarity between the reflectance of the crop and that of the water surface, the latter being influenced by the color of the tank, the wind profile over a crop differs from that over the tank, the heat storage in the tank can be considerable and the high temperature of the water increases the evaporation rate.

According to Chimene *et al.,* (2016) few studies have been carried out in Mozambique to evaluate the performance of different ETo estimation methods for irrigation management purposes. However, the conclusions vary greatly depending on the author's experimental conditions, which makes it difficult for the user to decide on the advisability of using a particular method. In view of the above, the aim of this research was to compare different empirical methods for estimating daily reference evapotranspiration in relation to ETo obtained by the Piche-INAM-Chimoio evaporimeter, estimated by the Hargreave-Samani method (1985), Oliveira and Carvalho (1998), in order to find the simplest method for estimating ETo and one that is accurate for irrigation management.

1.2. Study problem and justification

The irregularity of rainfall exposes agriculture to vulnerability, resulting in low yields. According to Viagem (2013), Mozambique needs to make the most of irrigation systems, sizing efficient systems to meet water needs at the right time. To this end, evapotranspiration is the main parameter used to quantify the amount of water lost from the soil surface and by the plant. Although estimating ETo is the main parameter used in estimating crop water requirements, in Mozambique there is a lack of research results in this area of study. According to Chimene et al. (2016), the study and analysis of evapotranspiration estimation in Mozambique is almost non-existent due to the unavailability of data. Since the method recommended by the FAO for estimating ETo requires various meteorological elements that are not always available (SOUSA, 2009). According to Camargo and Camargo (2000), when not all meteorological data is available, one alternative is to use simpler empirical methods with fewer input variables. According to Pilau *et al.* (2012) and Alencar *et al.* (2015), studies are recommended to assess the performance of simpler methods in estimating ETo for each region. In this sense, the use of different simple empirical methods that use a smaller number of meteorological variables in their formula was an alternative found for estimating ETo in this study. The research will contribute to increasing irrigated agriculture, planning the sowing season associated with crop production in the period of greatest atmospheric demand, improving cultural practices in the family sector, through correct water management in agriculture, avoiding applying too much or too little water, allowing for more efficient use of water resources since it will indicate the appropriate alternative methods for estimating reference evapotranspiration for the study site. In this context, it was important to carry out this study in order to compare different empirical methods for estimating reference evapotranspiration, in relation to ETo obtained by the Piche evaporimeter at the National Institute of Meteorology-Chimoio, estimated by the Hargreaves-Samani method (1985) and the Penman-Monteith method adjusted by Oliveira and Carvalho (1998), on a daily scale, using a six-month series of meteorological data, as a way of finding the simplest method for determining reference evapotranspiration and one that is accurate.

1.3. Hypothesis

There are empirical methods for estimating reference evapotranspiration that perform well as a function of the ETo obtained and estimated by the Piche-INAM-Chimoio evaporimeter, the Hargreaves-Samani (1985) and Penman-Monteith methods adjusted by Oliveira and Carvalho (1998), which were considered standard in this study.

1.4. Objectives

1.4.1. General

✓ Compare different empirical methods for estimating daily reference evapotranspiration in relation to ETo obtained by the Piche-INAM-Chimoio evaporimeter, estimated by the Hargreaves-Samani (1985) and Penman-Monteith methods adjusted by Oliveira and Carvalho (1998).

1.4.2. Specifics

✓ Estimate the daily reference evapotranspiration for each empirical method;

✓ Check the correlation between ETo obtained by each standard method and that estimated by the other empirical methods;

✓ Determine the performance of each empirical method in relation to the ETo estimation methods considered standard in this study.

II.   LITERATURE REVIEW

2.1.   Evapotranspiration

Knowledge of crop water requirements is of great importance when studying irrigation water management, as evapotranspiration is one of the main variables in the hydrological cycle and deserves careful control, otherwise it compromises crop performance and affects yields (SANTOS, 2015). The estimation of ETo is of great importance for various studies, such as hydrological balance, modeling of climatological processes, projects, irrigation management, simulation of crop productivity, planning and management of water resources (JÚNIOR *et al.,* (2012).

Allen *et al.* (1998) defined evapotranspiration as the combined effort of the evaporation of surface water or wetted surface, as well as the loss of water through plant transpiration. However, evapotranspiration was the term used by Thornthwaite in the early 1940s to express this simultaneous occurrence of evaporation and transpiration.

According to Netto and Bastos (2013), evapotranspiration is generally distinguished into three terms: (i) reference evapotranspiration · (ETo), (ii) crop evapotranspiration (ETc) and (iii) actual evapotranspiration (ETr). ETo is the evapotranspiration of a hypothetical low-growing 12cm crop, with an albedo of 0.23 and a surface resistance of 70 s/m, and resembles that of a surface covered in green grass, of uniform height, actively growing, with a completely shaded ground and no water restriction, and is dependent exclusively on the climatic conditions present in the location (BERNARDO *et al.,* 1996; NETTO and BASTOS, 2013). ETc is the evapotranspiration of a given crop when there are optimum conditions of humidity and nutrients in the soil, so as to allow the potential production of this crop in the field. Its determination is given by the product between ETo and the crop coefficient (Kc), which varies according to the type of crop and its phenological stage (STONE and SILVEIRA, 1995; CONCEIÇÃO, 2013 and FILHO *et al.,* 2010). ETr is that which occurs on the plant surface regardless of its area, size and soil water conditions (PEREIRA *et al.,* 1997).

2.2.   The importance of evapotranspiration

Agriculture is highly dependent on the climatic conditions prevailing in the growing regions and is significantly influenced by the various existing meteorological factors, including precipitation, solar radiation, wind and evapotranspiration (SILVA *et*

14

al., 2014). It is therefore important to study the influence of their variations on the different strategies for using the agricultural system, in order to provide subsidies for the decision-making process and optimize the planning of agricultural activities. According to Oliveira *et al* (2001), estimating evapotranspiration accurately contributes to the rational use of natural water resources and reduces production costs. The same authors point out that the importance of evapotranspiration is to quantify the actual amount of water to be supplied to the soil during crop irrigation.

According to Allen *et al* (1998), reference evapotranspiration (ETo) is used in agricultural water balances and in the modeling of climatological and hydrological processes, with the aim of estimating the need for irrigation, forecasting crops, assessing the availability of water resources and climate characterization.

According to Oliveira *et al.* (2001), knowledge of ETc is fundamental in irrigation projects, as it represents the amount of water that must be supplied to the soil in order to maintain growth and production under ideal conditions. Determining a crop's evapotranspiration throughout its development cycle is essential for estimating its water needs. According to Matos *et al.* (2015), knowing the water needs of crops during the development cycle, and in each sub-period, is fundamental for planning agricultural activity without the use of irrigation, improving the efficiency with which rainfall is used.

It is important to estimate evapotranspiration so that irrigation can be well quantified, thus avoiding insufficient watering that does not moisten the entire root zone, damaging the plants, wasting valuable resources and increasing the cost of applied water (Pereira *et al.* 1997).

2.3.   Factors that influence evapotranspiration

According to Carvalho *et al.* (2012) and Moura (2009), the main factors influencing evapotranspiration are climate, culture and soil.

2.3.1. Climatic factors

Solar radiation is the energy received on the earth's surface in the form of electromagnetic waves emitted by the sun. It is the main source of energy for the processes of photosynthesis in plants. The radiation falling on the earth's surface varies with altitude, climate and time of year, and is a fraction of the solar radiation at the top of the atmosphere (BORGES and MENDIONDO, 2007). According to Carlesso *et al.* (2007), the first relevant factor influenced by solar radiation is the heating of the air, causing the heating of materials and bodies on the earth's surface and consequently causing evaporation. It is worth noting that not all the energy available is used to evaporate water in the form of latent heat flow, but it is also used to heat the air in the form of sensible heat flow and the soil layers in the form of heat flow in the soil near the surface, and to a lesser extent for photosynthesis, heat storage by biomass (FILHO *et al.,* 2010).

Santos (2015) describes that air temperature is the main consequence of solar radiation. It is the main factor determining the natural distribution of plants. Regardless of the existence of favorable conditions of solar energy, water and nutrients in the soil, the plant growth of a given species ceases when the air temperature is below a minimum value or above a certain maximum value. The variation in air temperature over the course of a day determines the evaporative power of the air. According to Ribeiro (2006), an increase or decrease in air temperature leads to an increase or decrease in the air's water vapor saturation pressure deficit, which is directly proportional to the air's evaporative power. The same author states that the water saturation pressure deficit indicates the air's evaporation capacity and is given by the difference between the air's water vapor saturation pressure and the air's actual water vapor pressure. Dry air has a greater capacity to absorb additional water vapor than humid air, so as it approaches saturation, the evaporation rate decreases and tends to cancel itself out (PEREIRA *et al.,* 2004).

Relative humidity acts concomitantly with air temperature, but has an inverse relationship. The higher or lower the relative humidity, the lower or higher the evaporative power of the air, and therefore the lower or higher the evapotranspiration. According to Carlesso et al. (2007), wind affects plant growth in three ways: transpiration, CO absorption$_2$ and the mechanical effect on leaves and branches. Since wind is the movement of air in relation to the surface, this movement occurs when there is an atmospheric pressure gradient that moves the air mass from the point with the highest pressure to the points with the lowest pressure. The same author describes that when a moist air mass with a lower temperature is replaced by a dry and warmer air mass, new quantities of water vapor can be retained in the air, thus intensifying evaporation. The action of the wind is to displace the more humid air parcels located in the surface boundary layer, replacing them with drier ones. If there were no wind, the evapotranspiration process would cease as soon as the air reached saturation, since the atmosphere would be at its maximum capacity to absorb water vapor (SILVA et al., 2010).

2.3.2. Cultural Factors

Júnio *et al.* (2012) states that morphological aspects such as leaf architecture and the plant's internal resistance to water transport have a direct influence on evapotranspiration. The reflection coefficient for short waves (albedo) influences the availability of radiation for the evapotranspiration process. In practice, lighter surfaces reflect more solar radiation, and thus less available energy, compared to darker surfaces, which reflect less and absorb more. Therefore, darker vegetation, under the same climatic conditions, evapotranspires more than lighter vegetation.

The stage of crop development is related to the size of the transpiring leaf surface, represented by the leaf area index. The larger (smaller) the leaf area, the larger (smaller) the transpiring surface, increasing (decreasing) its water consumption (SANTOS, 2015).

The greater interaction of crops or larger plants with the atmosphere means that they experience greater evapotranspiration. The depth of the root system exploits a volume of soil that will supply a proportional amount of water to meet the plant's water consumption (PEREIRA et al. 2004).

2.3.3. Soil factors

According to Carvalho and Oliveira (2012), the soil is an active reservoir which, within certain limits, controls the rate at which plants consume water. This occurs in association with atmospheric water demand. In the hottest hours of the day, when atmospheric demand is high, the plant, even with soil moisture in the range of easily available water, is restricted from extracting water at a rate compatible with its needs. To prevent the leaves from wilting, the stomata close for a period of time (SANTOS, 2015).

On the other hand, the mineral composition of a soil interferes with its water storage capacity, with soils with a higher clay content having a higher storage capacity and soils with a lower clay content being able to compensate for the lower storage capacity if they are deeper and more permeable to the roots (CARVALHO and OLIVEIRA 2012; and SILVA *et al.*, 2014).

Evaporation is also influenced by soil cover, if the soil surface is very wet and is not shaded by plants or covered by the rest of the crops, the direct evaporation rate from the soil is comparable to the evaporation from a liquid surface (FERNANDES *et al.*, 2011). Moura (2009) points out that the contribution of soil evaporation to evapotranspiration decreases as vegetation cover increases. Therefore, when soil cover increases, evapotranspiration will depend on the availability of water in the soil and the suction capacity of the roots. However, there are situations near midday when, even with adequate water content in the soil, the transpiration rate decreases as a result of the increased resistance of the stomata to water diffusion.

According to Santos (2015), plant density determines intra-specific competition. Greater density results in intense competition for water, which causes the root system to deepen in order to increase water availability. The same author describes that lower density allows for greater heating of the soil and plants, and greater wind circulation between plants, and consequently an increase in evapotranspiration. Planting oriented perpendicular to the prevailing winds tends to extract more energy from the air than those oriented parallel (MANTOVANI *et al.*, 2013).

2.4. Methods for estimating evapotranspiration

According to Back (2008), the choice of method to be used depends on the availability of climate data. The methods available for calculating evapotranspiration involve meteorological variables that sometimes have a direct influence on its determination. For this reason, Moura (2009) pointed out that various methods can be used and tested according to the availability of existing data from weather stations and to check the compatibility of the results not only with the environment, but also with the influence of these variables on the formulated methodology.

Pereira *et al.* (1997) classified these methods into five categories: empirical, aerodynamic, energy balance, combined and eddy correlation. Among the various methods for estimating ETo on a daily scale, the FAO has adopted Penman-Monteith as the standard. This method, however, requires a greater amount of meteorological data, which often limits its use and is mainly used in research projects (NETTO and BASTOS, 2013).

According to Bernardo *et al.* (2013), the methods for determining ETo are divided into two groups, namely direct methods and indirect methods. According to Moura (2009), direct methods are those that use lysimeters, experimental plots in the field, soil moisture control and the method of water inflow and outflow over large areas. One of the direct methods for obtaining ETo is the use of a Piche evaporimeter (Figure 1), which consists of a calibrated glass tube closed at one end, filled with distilled water and closed with a circular filter paper attached with a spring, allowing consecutive measurements to determine evaporation over the desired period (PEREIRA *et al.*, 1997). The Piche evaporimeter measures the evaporative power of air in the shade on a daily basis. This equipment has very low acquisition costs compared to other meteorological equipment and, depending on environmental conditions, can satisfactorily determine ETo, helping with irrigation management (FERNANDES *et al.* 2011).

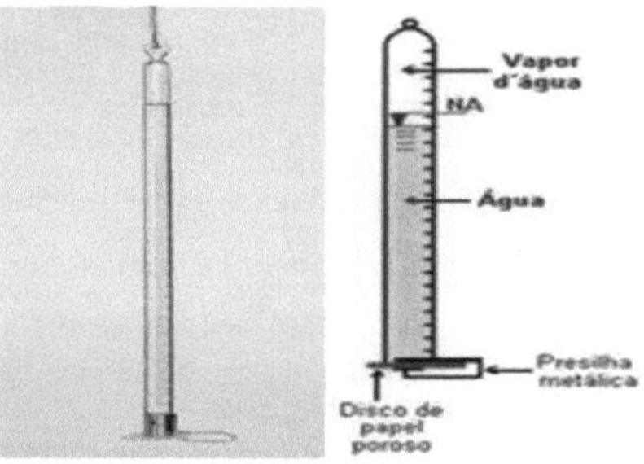

Figure 1Piche Evaporimeter Source: (GURSKI et al., 2016)

According to Carvalho & Oliveira (2012), the so-called indirect methods are classified as theoretical or empirical methods. Indirect methods estimate evapotranspiration using models that require knowledge of meteorological variables. These models can be simple, ranging from those based on temperature, such as Budyko, Holdrige, McCloud and others, to more complex ones involving the energy balance, such as Penman-Monteith (LORENZI, 2010). The evapotranspiration of a specific crop can be estimated from ETo. On the other hand, Allen *et al.* (1998) state that among the indirect methods, the Penman-Monteith method, recommended by the FAO, is considered the standard for estimating reference evapotranspiration. The Penman-Monteith method was adopted as the most suitable for determining ETo on a daily scale at a meeting organized by the FAO which brought together a group of evapotranspiration experts in 1990, and is recommended as the standard for comparison with other methods (PEREIRA *et al.*, 1997 and OLIVEIRA *et al.*, 2001). Also according to the authors, the Penman-Monteith method is recommended by the FAO for determining reference evapotranspiration, since its value is very close to the ETo of the grass at its evaluation site, has a physical basis and incorporates both physiological and aerodynamic parameters.

The Penman-Monteith method uses meteorological variables such as air temperature, relative humidity, solar radiation balance, soil heat flow, wind speed, vapor pressure and saturation pressure (BACK 2008). However, this method requires input data that makes it difficult to apply, since these meteorological elements are not always available in some regions (FILHO *et al.*, 2010 and NETTOS and BASTOS, 2013).

According to Mendonça (2003), when choosing the method to be used to estimate ETo, it is necessary to know the climatic elements available, the level of precision required and the purpose. As an alternative to solving this problem, we highlight the use of empirical methods for estimating reference evapotranspiration, which, because they are developed and calibrated locally, cannot be applied universally, but which present better results than those that are more generic and physically more real (RIBEIRO, 2016).

2.5.   Evaluation of ETo estimation methods

According to Back (2008), the use and recommendation of an empirical method for estimating reference evapotranspiration should be based on its accuracy, practicality and the availability of meteorological data; to this end, it is necessary to evaluate and calibrate the various methods to local climatic conditions. Oliveira *et al* (2001) compared evapotranspiration estimation methods for some locations in the state of Goiás and the Federal District. The Hargreave Samani method (1985) showed high significance with the standard Penman-Monteith method.

Junior *et al.*, (2011), with the aim of evaluating 15 methods for estimating ETo at the Federal Rural University of the Semi-Arid in conditions of low (dry) and high (wet) relative humidity using the Penman Monteith-FAO 56 method as a reference, observed that the Benevides Lopes method was the one that obtained optimum performance for both periods. The Hargreave Samani 1985 and Benevidez Lopez methods performed well in the wet period, and the Blaney-Criddle FAO24 method performed well in both the wet and dry periods. In the same study, the Camargo and Hamon model performed very poorly in the dry period.

In the study by Macuiza (2018), he concluded that the McCloud, Oliveira and Carvalho (1998) and Benevides-Lopes methods showed good accuracy and correlation, "very good" and good performance on a daily scale, considering ETo obtained by the Hargreaves-Samani (1985) method as the standard. Considering the method of Oliveira and Carvalho (1998) as the standard, he concluded that Benevides-Lopes, Hargreaves-Samani (1985) and modified Hargreaves showed good precision and correlation, and very good and good performance on the daily scale.

Mendonça et al (2003) compared the evapotranspiration values obtained in a weighing lysimeter with grass with the values resulting from the use of 6 empirical ETo estimation methods. The authors concluded that the methods of Linacre and Hargreave Samani (1985) were satisfactory for estimating evapotranspiration in the Norte Fluminense region of Rio de Janeiro. The authors also concluded that in the absence of data on climatic variables, the Hargreaves method is accurate for estimating evapotranspiration values for the region.

III. MATERIALS AND METHODS S

3.1. Data connection

The data used to estimate ET_0 was obtained from the Meteorological Station of the National Institute of Meteorology (INAM), Manica Province Delegation, Chimoio City, located in the Ttrangapasso neighborhood (Figure 2). Daily data was collected on air temperature (minimum, maximum and average), average relative humidity, average wind speed and ETo obtained using a Piche evaporimeter. The meteorological data was collected over a six-month period (October 2019 to March 2020) corresponding to 183 days, according to the methodology used by Silva *et al.* (2015) and Macuza (2018). This period was chosen because it is the growing season for the main summer crops that constitute food security in Mozambique. These crops are more susceptible to water deficiency when rainfall is irregular, making it necessary to use supplementary irrigation to avoid crop losses (CHIMENE, 2016). Some of the data used to estimate ETo depends on the latitude and time of year, such as the number of hours of sunshine (N), percentage of maximum daily insolation (p) and solar radiation at the top of the atmosphere (Ro) (Table 1).

Figure 2Meteorological Station of the National Meteorological Institute (INAM), Manica Province Delegation, Chimoio City. Source: Author (2020).

Table 1Monthly values corresponding to the number of hours of sunshine (N), percentage of maximum daily insolation (p) on the 15th of each month, solar radiation at the top of the atmosphere (Ro) for the period under study from October to March

Month	N(horas)	P	Ro(MJ/m^2/dia)	Ro (mm/dia)
October	12.5	0.28	38.6	15.3
November	13.2	0.29	41.2	16.4
December	13.3	0.30	42.1	16.8
January	13.2	0.30	41.9	16.7
February	12.8	0.29	40.0	16.0
March	12.2	0.28	36.6	14.5

Source: Pereira *et al.* (1997) and Bernardo et *al.* (2013)

3.2. Determining Reference Evapotranspiration

Thirty (30) empirical methods were used to estimate daily ETo, namely: Piché-INAM, Hargraves-Samani (1985) (equation 1), Penman-Montheith adjusted by Oliveira and Carvalho (1998) (equation 2), Budyko (1956) (equation 3), Holdrige (equation 4), Blaney-Morin (equation 5), modified Hargraves (equation 6), Ivanov (1977) (equation 7), McGuiness-Bordne (2005) (equation 8), Camargo (1971) (equation 9), Hamon (1961) (equation 10), Kharrufa (1985) (equation 11), Benevides-Lopez (1970) (equation 12), McCloud (2001) equation 13), Romanenko modified by Oudth *et al.,* (2005) (equation 14), Shendel (1967) (equation 15), FAO-Blaney-Cride (1950) (equation 16), Linacre (1977) (equation 17), Trajkovic (2007) (equation 18), Ravazzani *et al.* (2012) (equation 19), Dalton (equation 20), Trabert (1896) (equation 21), Meyer (1926) (equation 22), Rhwer (1931) (equation 23), Penman (1948) (equation 24), Albrecht (1950) (equation 25), Brokamp et al (1963) (equation 26), WMO (1966) (equation 27), Mahringer (1970) (equation 28) and Hargreaves Modified by Bert *et al* (2014) (equation 29). The methods of Piche-INAM, Hargraves-Samani (1985) and Penman-Montheith adjusted by Oliveira and Carvalho (1998) were considered standard.

3.3. Methods for Estimating Reference Evapotranspiration

The different ETo estimation methods used in this research were described by Allen *et al.* (1998), Borges and Mendiondo (2007), Marcuzzo et al. (2008), Mendonça

(2003), Back (2008), Júnior et al. *(2011)*, Júnior et *al.* *(2012)*, Fernandes *et al. (*2011), Carvalho and Oliveira (2012), Bernardo et al, (2013), Silva *et al.,* (2014), Alencar et al. (2015), Macuza (2018), Djaman et *al.,* (2015), Matos e Silva (2015), Santana et *al., (2018),* Lopes *et al., (2018),* Rincón (2018), Muhammad *et al.,* (2019) and Brandão (2019), as can be seen in the following expressions:

3.3.1. Hargreaves-Samani method (1985)

$$ETo = 0,0023Ro\sqrt{T_{max} - T_{min}} * (T_{med} + 17,8) \qquad \text{(Equation 1)}$$

Where:

ETo = Reference evapotranspiration (mm day)[-1];

Ro = incident solar radiation at the top of the atmosphere (MJ mm^{-2} day)[-1];

T_{max} = maximum air temperature (°C);

T_{min} = minimum air temperature (°C);

T_{med} = average air temperature (°C).

3.3.2. Penman-Monteith method adjusted by Oliveira and Carvalho (1998).

$$ETo(P - M) = -0,023 + 0,8161 * ETo(HG) \qquad \text{(Equation 2)}$$

Where:

ETo (P-M) = Adjusted P-M reference evapotranspiration (mm day)[-1];

ETo (HG) = Hargreaves-Samani reference evapotranspiration (mm day)[-1].

3.3.3. Budyko method (1956)

$$ETo = 0,2 * T_{med} \qquad \text{(Equation 3)}$$

Where:

ETo = reference evapotranspiration (mm day)$^{-1}$;

T_{med} = Average air temperature ($^{\circ}$ C).

3.3.4. Holdrige method

$$ETo = CHO * T_{med} \qquad \text{(Equation 4)}$$

Where:

ETo = reference evapotranspiration (mm day)$^{-1}$;

CHO = 0.161 for daily evapotranspiration estimates;

T_{med} = Average air temperature ($^{\circ}$C).

3.3.5. Blaney-Morin method

$$ETo = p(0{,}457 T_{med} + 8{,}13) * (1{,}14 - 0{,}01H) \qquad \text{(Equation 5)}$$

Where:

ETo = Reference evapotranspiration (mm day)$^{-1}$;

p = monthly percentage of light (%);

T_{med} = Average air temperature ($^{\circ}$C);

H = average relative humidity (%).

3.3.6. Modified Hargreaves method

$$ETo = Ro(1{,}8 * T_{med} + 32) * 0{,}0006 * \sqrt{100 - H} \qquad \text{(Equation 6)}$$

Where:

ETo = Reference evapotranspiration (mm day); [-1]

Ro = incident solar radiation at the top of the atmosphere (KJ mm^{-2} day); [-1]

H = average relative humidity (%);

T = average air temperature (°C).

3.3.7. Ivanov's method (1977)

$$ETo = 0{,}006 * (25 + T_{med})^2 * \left(1 - \frac{H}{100}\right)$$ (Equation 7)

Where:

ETo = reference evapotranspiration (mm day); [-1]

T_{med} = average air temperature (°C);

H = average relative humidity (%).

3.3.8. McGuiness-Bordne method (2005)

$$ETo = \frac{Ro}{\lambda} * \frac{T_{med} + 5}{68}$$ (Equation 8)

Where:

ETo = reference evapotranspiration (mm day); [-1]

Ro = incident solar radiation at the top of the atmosphere (MJ m^{-2} day); [-1]

T_{med} = average air temperature (°C);

λ = latent heat of vaporization (2.45 MJ kg-1).

3.3.9. Camargo's method (1971)

$$ETo = 0{,}01 * \left(\frac{Ro}{2{,}45}\right) * T_{med}$$ (Equation 9)

Where:

ETo = reference evapotranspiration (mm day-1);

Ro = incident solar radiation at the top of the atmosphere (MJ m^{-2} day)$^{-1}$;

T_{med} = Average daily air temperature (ºC).

3.3.10. Hamon's method (1961)

$$ETo_{(HA)} = 0,55 * \left(\frac{N}{12}\right)^2 * \left(\frac{4,93^{(0,062*T_{med})}}{100}\right) * 25,4 \qquad \text{(Equation 10)}$$

Where:

ETo (HA) = reference evapotranspiration (mm day)$^{-1}$;

N = number of possible hours of sunlight (photoperiod);

T_{med} = Average air temperature (°C).

3.3.11. Kharrufa method (1985)

$$ETo = 0,34 * p * T_{med}^{1,3} \qquad \text{(Equation 11)}$$

Where:

ETo is the reference evapotranspiration (mm day)$^{-1}$;

p = percentage of maximum daily sunshine (N) in relation to sunshine hours;

T_{med} = Average air temperature (°C).

3.3.12. Benevides-Lopez method (1970)

$$ETo = 1,21 * 10^{\left[7,5*\frac{T_{med}}{237,5+T_{med}}\right]} * (1 - 0,01 * H) + 0,21 * T_{med} - 2,3 \qquad \text{(Equation 12)}$$

Where:

ETo = reference evapotranspiration (mm day);$^{-1}$

T_{med} = Average air temperature (°C);

H = Average relative humidity (%).

3.3.13. McCloud's method (2001)

$$ETo = 0{,}254 * 1{,}07^{(1{,}8*T_{med})}$$ (Equation 13)

Where:

ETo = reference evapotranspiration (mm day); $^{-1}$

T_{med} = Average air temperature (°C).

3.3.14. Romanenko's method modified by Oudth et al (2005)

$$ETo = 4{,}5 * \left(1 + \frac{T_{med}}{25}\right)^2 * \left(1 - \frac{ea}{es}\right)$$ (Equation 14)

$$e_s = 0.6108 * \exp^{\frac{17{,}3*T_{med}}{237{,}3+T_{med}}}$$

$$e_a = \frac{H * e_s}{100}$$

Where:

ETo = reference evapotranspiration (mm day); $^{-1}$

T_{med} = Average daily air temperature (°C);

e_s = Vapor saturation pressure (hPa);

e_a = Vapor pressure (hPa);

H = average relative humidity (%).

3.3.15. Schendel's method (1967)

$$ETo = 16 * \left(\frac{T_{med}}{H}\right) \qquad \text{(Equation 15)}$$

Where,

ETo = reference evapotranspiration (mm day); $^{-1}$

T_{med}= Maximum daily air temperature (°C);

H= relative humidity (%);

3.3.16. FAO method Blaney-Criddle (1950)

$$ETo = p * (0{,}46T_{med} + 8{,}13) \qquad \text{(Equation 16)}$$

Where:

ETo = reference evapotranspiration (mm day); $^{-1}$

T_{med}= Maximum daily air temperature (°C);

p = percentage of maximum daily sunshine (N) in relation to sunshine hours;

3.3.17. Linacre's method (1977)

$$ETo = \frac{\left[\frac{700*(T+0.006*h)}{100-lat}\right] + 15 * (T_{med} - Td)}{80 - T_{med}} \qquad \text{(Equação 17)}$$

Where:

ETo = reference evapotranspiration (mm day); $^{-1}$

Td - Dew point temperature (°C);

$$Td = \frac{[273.3 * \ln(e_a) - 429.41]}{[19.078955 - \ln(e_a)]}$$

lat - Latitude (°) ;

T_{med} = Maximum daily air temperature (°C); and

h - altitude (m).

3.3.18. Trajkovic's method (2007)

$$Eto = (0,0023 * R_0) * Td^{0.424} * (T_{med} + 17.8) \qquad \text{(Equation 18)}$$

Where:

ETo = reference evapotranspiration (mm day); [-1]

Ro = incident solar radiation at the top of the atmosphere (MJ m[-2] day); [-1]

T_{med} = Maximum daily air temperature (°C);

Td - Dew point temperature (°C),

3.3.19. Method by Ravazzani et al. (2012)

$$ETo = (0,817 + 0,00022h)(0,0023R_0)(Td^{0.5})(T_{med} + 17.8) \qquad \text{(Equation 19)}$$

Where:

ETo = reference evapotranspiration (mm day); [-1]

T_{med} = Maximum daily air temperature (°C);

Td - Dew point temperature (°C),

h - Altitude (m).

Ro = incident solar radiation at the top of the atmosphere (MJ m-2 day-1);

3.3.20. Dalton's method (1802)

$$\text{ETo} = (0{,}3648 + 0{,}07223 * u_2) * (e_s - e_a) \qquad \text{(Equation 20)}$$

Where:

ETo = reference evapotranspiration (mm day)[-1];

es = vapor saturation pressure (Pa);

ea = vapor pressure (Pa),

u_2 = Average air speed measured at a height of 2 meters in relation to the surface given by the following equation.

$$u_2 = u_z * \frac{4.87}{\ln(67{,}8 * z - 5{,}42)}$$

Where:

u_2 = Wind speed measured at a height of 2 meters in m/s,

u_z = Speed measured at the height of the instrument at the station (m/s),

z = the height of the instrument installation (m).

3.3.21. Trabert's method (1896)

$$\text{ETo} = (0{,}3075) * \sqrt{u_2} * (e_s - e_a) \qquad \text{(Equation 21)}$$

Where:

ETo = reference evapotranspiration (mm day)[-1];

u_2 = Average air speed measured at a height of 2 meters from the surface,

es = saturation vapor pressure (hPa);

ea = vapor pressure (hPa).

3.3.22. Meyer's method (1926)

$$ETo = (0{,}375 + 0{,}05026 * (u_2) * (e_s - e_a)$$ (Equation 22)

Where:

ETo = reference evapotranspiration (mm day); $^{-1}$

u_2 = Average air speed measured at a height of 2 meters from the surface,

es = saturation vapor pressure (hPa);

ea = vapor pressure (hPa).

3.3.23. Rohwer's method (1931)

$$ETo = (3{,}3 + 0{,}0891 * u_2) * (e_s - e_a)$$ (Equation 23)

Where:

ETo = reference evapotranspiration (mm day); $^{-1}$

u_2 = average air speed measured at a height of 2 meters from the surface,

es = vapor saturation pressure (Pa);

ea = vapor pressure (Pa).

3.3.24. Penman method (1948)

$$ETo = \left(2{,}625 + \frac{0{,}000479}{u}\right) * (e_s - e_a)$$ (Equation 24)

Where:

ETo = reference evapotranspiration (mm day); $^{-1}$

u_2 = Average air speed measured at a height of 2 meters from the surface, es = saturation vapor pressure (hPa);

ea = vapor pressure (hPa).

3.3.25. Albrecht's method (1950)

$$ETo = (0{,}1005 + 0{,}297 * u_2) * (e_s - e_a) \qquad \text{(Equation 25)}$$

Where:

ETo = reference evapotranspiration (mm day)$^{-1}$;

u_2 = Average air speed measured at a height of 2 meters from the surface,

es = saturation vapor pressure (hPa);

ea = vapor pressure (hPa).

3.3.26. Brokamp et al method (1963)

$$ETo = 0{,}543 * (u_2{}^{0.456}) * (e_s - e_a) \qquad \text{(Equation 26)}$$

Where:

ETo = reference evapotranspiration (mm day)$^{-1}$;

u_2 = Average air speed measured at a height of 2 meters from the surface,

es = saturation vapor pressure (hPa);

ea = vapor pressure (hPa).

3.3.27. World Meteorological Organization (WMO) method (1966)

$$ETo = (0{,}1298 + 0{,}0934 * u_2) * (e_s - e_a) \qquad \text{(Equation 27)}$$

Where:

ETo = reference evapotranspiration (mm day); $^{-1}$

u_2 = average air speed measured 2 meters above the surface (m/s),

es = saturation vapor pressure (hPa);

ea = vapor pressure (hPa).

3.3.28. Mahringer's method (1970)

$$ETo = (0,15072) * \sqrt{3 * 6 * u_2} * (e_s - e_a)$$ (Equation 28)

Where:

ETo = reference evapotranspiration (mm day); $^{-1}$

u_2 = Average air speed measured at a height of 2 meters from the surface,

es = saturation vapor pressure (hPa);

ea = vapor pressure (hPa).

3.3.29. Hargreaves method modified by Bert et. all (2014)

$$ETo = 0.00193 * Ro * (T_{med} + 17.8) * (T_{max} - T_{min})^{0.517}$$ (Equation 29)

Where:

ETo = reference evapotranspiration (mm day); $^{-1}$

Ro = incident solar radiation at the top of the atmosphere (mm/day);

T_{med}= Average daily air temperature (°C);

T_{max}= Maximum daily air temperature (°C);

T_{min}= Minimum daily air temperature (°C);

3.4.   Comparison of ETo Estimation Methods

The daily ETo estimated by each method was compared with the ETo obtained by the Piche evaporimeter provided by INAM, with the ETo estimated by the Hargreaves-

Samani method (1985) (equation 1) and Penman-Monteith adjusted by Oliveira and Carvalho (1998) (equation 2), according to the methodologies used by Macuiza, (2018). The Hargreaves-Samani method (1985) is recommended by the FAO as an alternative when there is not enough data to feed the Penman-Monteith FAO-56 method. On the other hand, the Penman-Monteith method adjusted by Oliveira and Carvalho (1998) is proposed by Carvalho and Oliveira (2012) to estimate ETo, obtaining values similar to those of the Penman-Monteith method recommended by the FAO, and can be used to quantify the correct supplementary irrigation blade in agricultural production fields.

3.5.   Data Analysis

To validate the linear regression coefficients, the hypotheses $H_0 : a = 0$ versus Ha: $a \neq 0$ and $H_0 : b = 1$ versus Ha: $b \neq 1$ were tested using Student's t-test at 5%, using Microsoft Office Excel® (2010) software (appendix 1, 2 and 3).

The results were analyzed by comparing the ETo values obtained by the other empirical methods (Xi) with the standard methods (Yi) Penman-Monteith, Hargrave-Samani and Piche ETo. Linear regression analysis was carried out to obtain the coefficients of the equation $Yi = a + bXi$ (where "a" and "b" are linear regression coefficients), the coefficient of determination (R^2) and correlation coefficient (r).

To analyze the accuracy of the methods evaluated (Xi) and the standard method (Yi), the mean absolute error (MAS) (equation 34), root mean square error (RQME) (equation 35), coefficient of determination (R^2), correlation coefficient (r) (equation 36) and agreement index (d) of Willmott *et al.* (1985) (equation 37), whose values vary from 0.0 for no agreement to 1.0 for perfect agreement, according to the recommendations given by Camargo and Camargo (2000). The CS index of Camargo and Sentelhas (1997) was obtained from the product of the correlation coefficient (r) and the agreement index (d) of Willmott *et al.* (1985), as shown in equation (38).

$$EAM = \frac{\sum_{i=1}^{n} |X_i - Y_i|}{n} \qquad \text{(Equaçã 34)}$$

$$RQME = \sqrt{\frac{\sum_{i=1}^{N}(X_i - Y_i)^2}{n}}$$ (Equação 35)

Where: EAM = erro absoluto médio;

RQME = raiz quadrado médio do erro;

X_i = valores de ETo estimado pelos demais metodos (mm/dia);

Y_i = valores de ETo estimado por método padrão (mm/dia);

X_{med} = media dos valores estimado de ETo por demais metodos $\left(\frac{mm}{dia}\right)$;

Y_{med} = média dos valores de ETo estimados pelo método padrão (mm/dia);

n = número de dias (183 dias).

$$r = \frac{\sum_{i=1}^{n}(Y_i - Y_{med}) * (X_i - X_{med})}{\sqrt{\sum_{i=1}^{n}(Y_i - Y_{med})^2} * \sqrt{\sum_{i=1}^{n}(X_i - X_{med})^2}}$$ (Equação 36)

$$d = 1 - \left[\frac{\sum_{i=1}^{N}(X_i - Y_i)^2}{\sum_{i=1}^{N}[(X_i - Y_{med}) + (Y_i - Y_{med})]^2}\right]; \ 0 \leq d \leq 1$$ (Equação 37)

Where: r = coefficient de correlação;

d = indice de concordância de Willmott et al. (1985);

Y_i = valores de ETo estimado por método padrão (mm/dia);

X_i = valores de ETo estimado pelos demais metodos (mm/dia);

X_{med} = media dos valores estimado de ETo por demais metodos $\left(\frac{mm}{dia}\right)$;

Y_{med} = média dos valores de ETo estimados pelo método padrão (mm/dia);

n = número de dias (183 dias);

$$CS = r * d$$ (Equação 38)

Where: CS = indice de desempenho;

The criteria for interpreting the performance of the CS index of Camargo and Sentelhas (1997) and classifying the correlation coefficient of the methods are shown in Table 2.

Table 2: Classification of the correlation coefficient (r) and the performance of reference evapotranspiration (ETo) methods, by the CS index of Camargo and Sentelhas (1997).

CS value	Performance	Correlation Coefficient (r)	Correlation
>0.85	Great	0.9-1.0	Almost perfect
0.76-0.85	Very good	0.7-0.9	Very high
0.66-0.75	Good	0.5-0.7	High
0.61-0.65	Average	0.3-0.5	Moderate
0.51-0.60	Sufferable	0.1-0.3	Low
0.41-0.5	Bad	0.0-0.1	Very low
≤ 0.40	Bad		

Source: Sousa *et al.*, (2009), Camargo and Sentelhas (1997).

3.6.   ETo Estimation Method Selection Criteria

The criteria used to compare the empirical methods that best estimate ETo in relation to the standard method were: angular coefficient (a) closest to one, linear coefficient (b) closest to zero, Pearson's linear correlation coefficient (r) and coefficient of determination (R^2) closest to one, mean absolute error (MAS) and root mean square error (RQME) closest to zero, and index of agreement (d) by Willmott *et al*, (1985) and the CS index of Camargo and Sentelhas (1997) closer to one.

IV. RESULTS AND DISCUSSION

4.1. Weather conditions: Temperature, relative humidity and wind speed

Figure 3 shows the meteorological variables air temperature (minimum, maximum and average), relative humidity (average) and average wind speed, which showed a daily variation in data during the period under analysis (183 days). The maximum air temperature ranged from 17.1 to 40.9 °C, the minimum temperature ranged from 14.8 to 23.8 °C and the average temperature ranged from 15.6 to 32.2 °C, with the extreme values being observed in October (Figure 3A). The average relative humidity varied from 36 to 100 %, reaching a maximum in October and a minimum in February (Figure 3B), the opposite of the air temperature. Filho *et al.,* (2010), Oliveira *et al.,* (2001) and Tanaka et *al.,* (2016) found the same result, where temperature behaved inversely to relative humidity. The average wind speed ranged from 2.1 to 20.6 m/s, with the lowest value in October and the highest in February (Figure 3C).

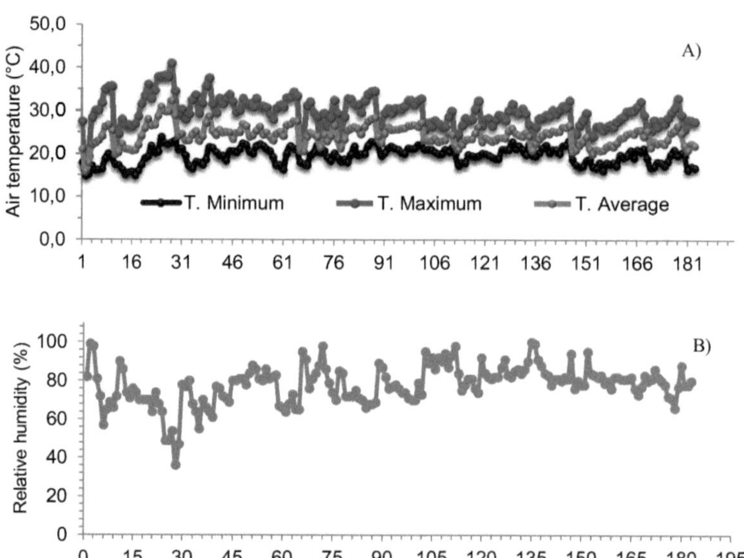

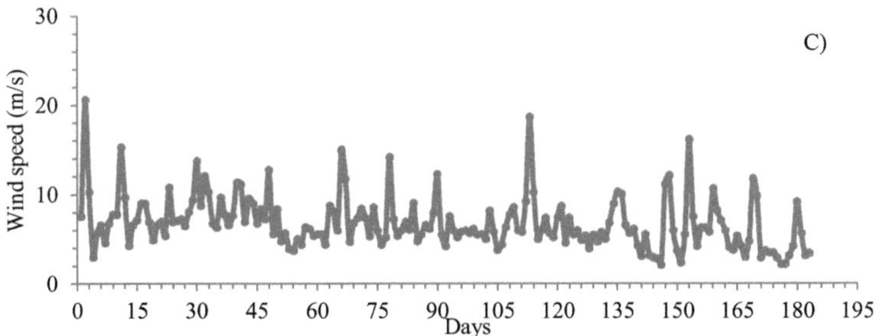

Figure 3Daily variation in meteorological data: A) air temperature (maximum, minimum and average) (°C), B) relative humidity (average) (%) and wind speed (average) (m/s), from October 2019 to March 2020. Source: Author 2020

4.2. Evaluation of ETo estimation methods using ETo obtained by the Piche-INAM method as a standard.

The evaluation of the ETo estimation methods of Hargreaves-Samani (1985), Penman-Monteith adjusted by Oliveira and Carvalho (1998), Budyko (1956), Holdrige, Blaney-Morin, modified Hargreave, Ivanov (1977), McGuiness-Bordne (2005), Camargo (1971), Hamon (1961), Kharufa (1985), Benevides-Lopes (1970), McCloud (2001), Romanenko Modified by Oudth et al (2005), Shendel (1967), Blaney-Criddle (1950), Linacre (1977), Trajkovic (2007), Ravazzani (2012), Dalton (1802), Trabert (1896), Meyer (1926), Rohver (1931), Penman (1948), Albrecht (1950), Brokamp et al. (1963), World Meteorological Organization (1966), Mahringer (1970) and Hargreaves de Bert et al. (2014) considering ETo estimated by the Piche-INAM method as a standard, is shown in Figure 4.

The values of daily ETo, the errors (EAM and RQME), Pearson's correlation coefficient "r" and determination coefficient "R^2 ", angular coefficient "a" and linear coefficient "b", agreement index "d" by Willmott et al. (1985) and performance index "CS" by Camargo and Sentelhas (1997) of each method in relation to ETo-INAM are shown in Table 3.

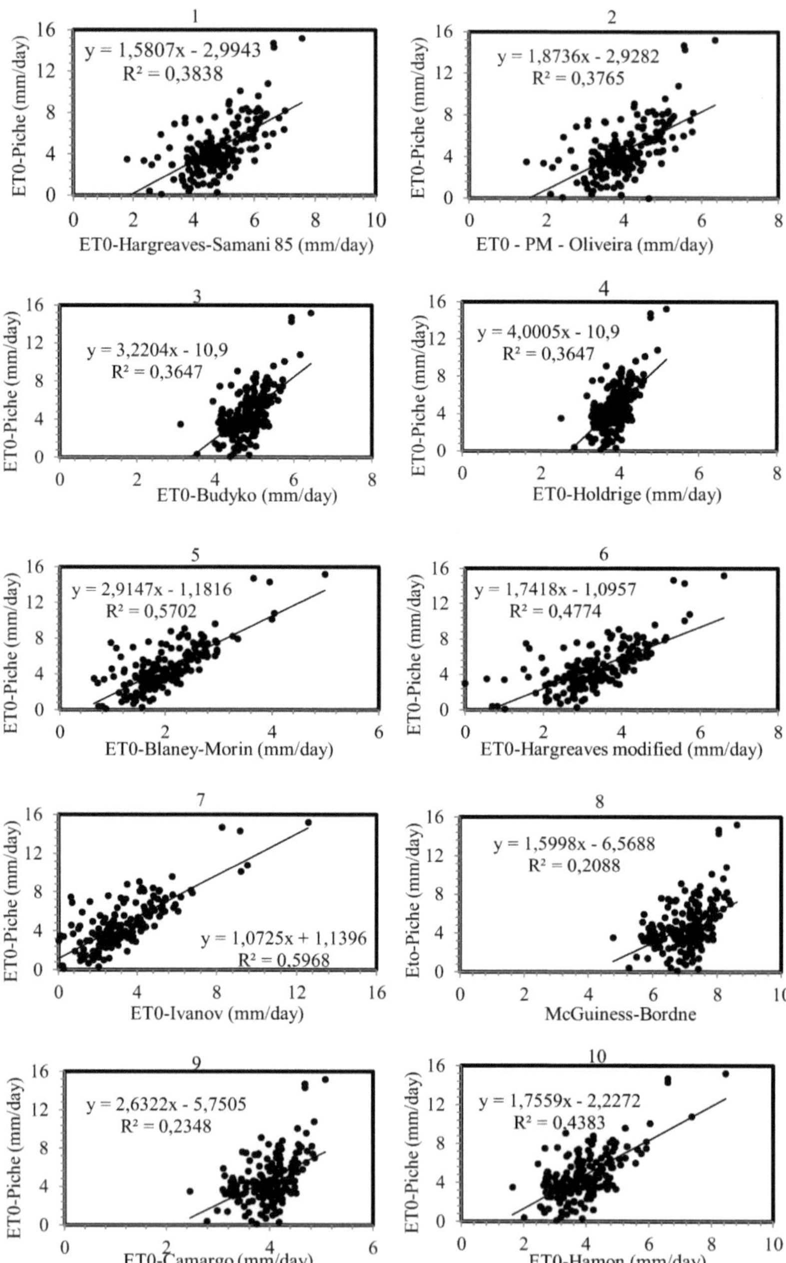

1
y = 1,5807x - 2,9943
R² = 0,3838
ET0-Piche (mm/day)
ET0-Hargreaves-Samani 85 (mm/day)

2
y = 1,8736x - 2,9282
R² = 0,3765
ET0-Piche (mm/day)
ET0 - PM - Oliveira (mm/day)

3
y = 3,2204x - 10,9
R² = 0,3647
ET0-Piche (mm/day)
ET0-Budyko (mm/day)

4
y = 4,0005x - 10,9
R² = 0,3647
ET0-Piche (mm/day)
ET0-Holdrige (mm/day)

5
y = 2,9147x - 1,1816
R² = 0,5702
ET0-Piche (mm/day)
ET0-Blaney-Morin (mm/day)

6
y = 1,7418x - 1,0957
R² = 0,4774
ET0-Piche (mm/day)
ET0-Hargreaves modified (mm/day)

7
y = 1,0725x + 1,1396
R² = 0,5968
ET0-Piche (mm/day)
ET0-Ivanov (mm/day)

8
y = 1,5998x - 6,5688
R² = 0,2088
Eto-Piche (mm/day)
McGuiness-Bordne

9
y = 2,6322x - 5,7505
R² = 0,2348
ET0-Piche (mm/day)
ET0-Camargo (mm/day)

10
y = 1,7559x - 2,2272
R² = 0,4383
ET0-Piche (mm/day)
ET0-Hamon (mm/day)

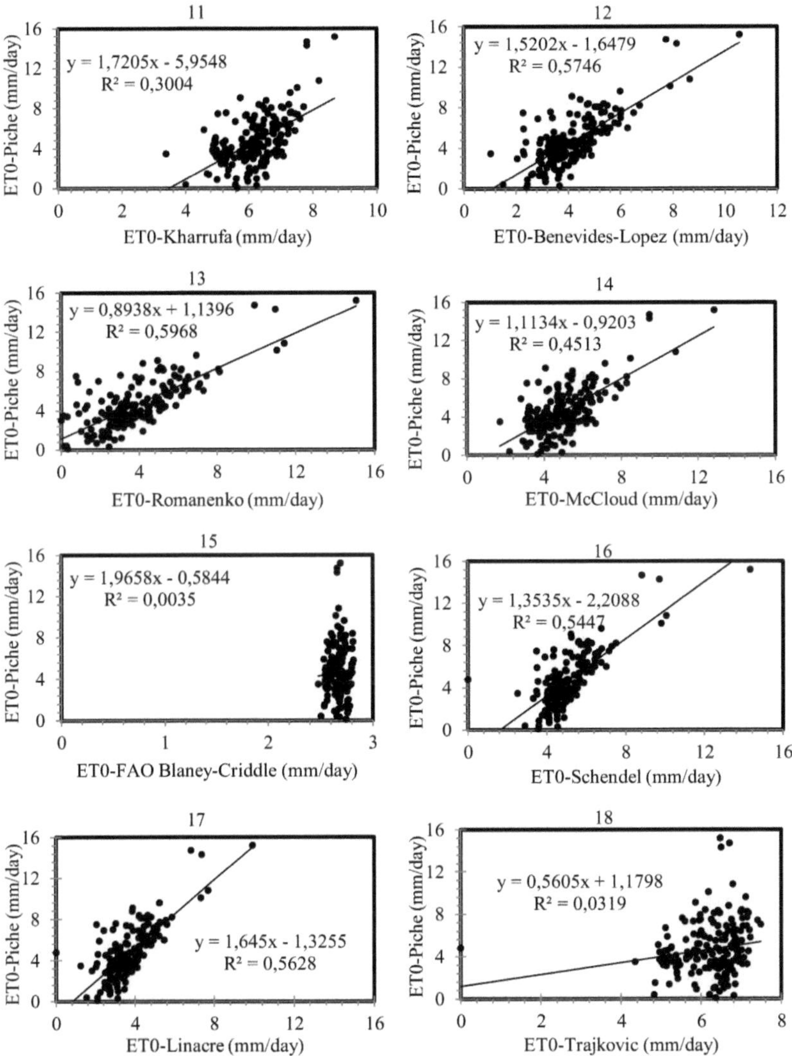

41

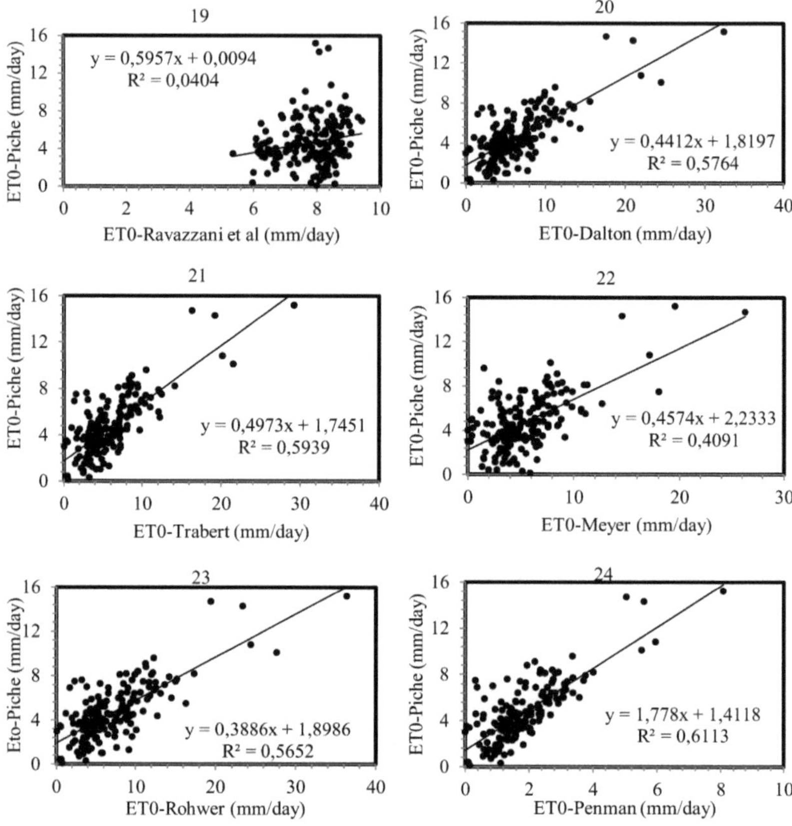

19
y = 0,5957x + 0,0094
R² = 0,0404
ET0-Piche (mm/day)
ET0-Ravazzani et al (mm/day)
20
y = 0,4412x + 1,8197
R² = 0,5764
ET0-Piche (mm/day)
ET0-Dalton (mm/day)
21
y = 0,4973x + 1,7451
R² = 0,5939
ET0-Piche (mm/day)
ET0-Trabert (mm/day)
22
y = 0,4574x + 2,2333
R² = 0,4091
ET0-Piche (mm/day)
ET0-Meyer (mm/day)
23
y = 0,3886x + 1,8986
R² = 0,5652
Eto-Piche (mm/day)
ET0-Rohwer (mm/day)
24
y = 1,778x + 1,4118
R² = 0,6113
ET0-Piche (mm/day)
ET0-Penman (mm/day)

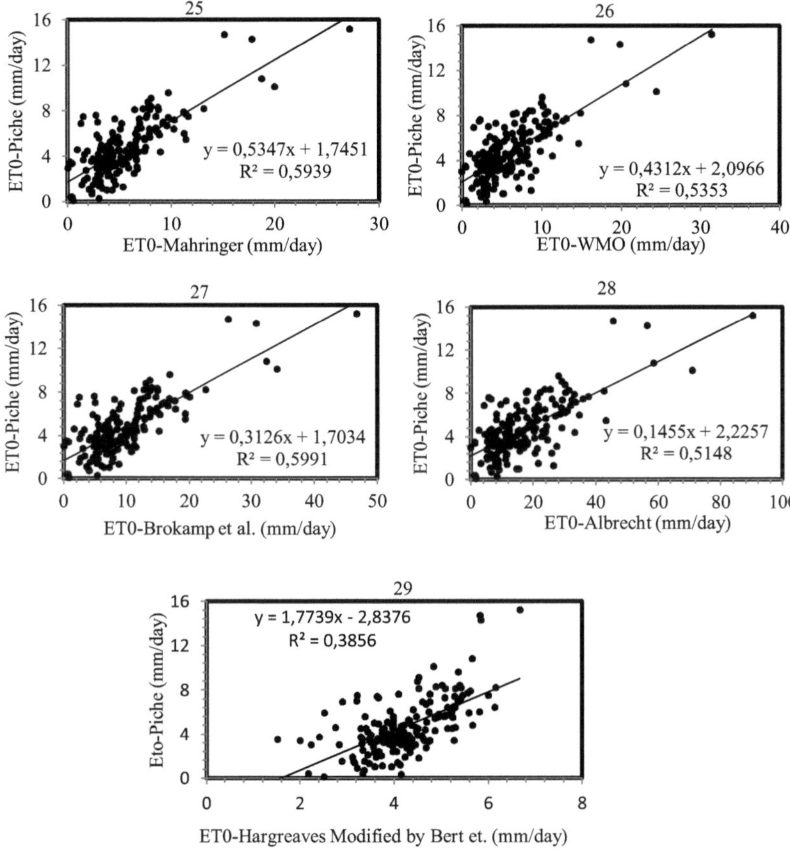

Figure 4: Linear regression of ETo values estimated by the Piche method, compared with ETo values estimated by the Hargreaves-Samani (1985), Penman-Monteith adjusted by Oliveira and Carvalho (1998), Budyko (1956), Holdrige, Blaney-Morin, modified Hargreave, Ivanov (1977), McGuiness-Bordne (2005), Camargo (1971), Hamon (1961), Kharufa (1985), Benevides-Lopes (1970), McCloud (2001), Romanenko Modified by Oudth et al (2005), Shendel (1967), Blaney-Criddle (1950), Linacre (1977), Trajkovic (2007), Ravazzani (2012), Dalton (1802), Trabert (1896), Meyer (1926), Rohver (1931), Penman (1948), Albrecht (1950), Brokamp et al. (1963), World Meteorological Organization (1966), Mahringer (1970) and Hargreaves de Bert et al. (2014), for 183 days, from October 1, 2019 to March 31, 2020. Source: Author (2020).

Table 3: ETo (mm day^{-1}) of different methods, EAM, RQME, Pearson's linear correlation coefficient (r), linear (a) and angular (b) coefficients obtained in the linear regression between ETo obtained by the Piche-INAM method (mm) (dependent variable) and by different methods (independent variable). Accuracy index (d) from Willmott *et al.* (1985) and CS index from Camargo and Sentelhas (1997) and person's correlation coefficient (r) calculated based on the estimated ETo (mm.day^{-1}), from October 2019 to March 2020. Source: Author (2020).

Method	ETo	EAM	RQME	a[1]	b[2]	r	d	CS	Performance[3]	Correlation[4]
ETo-Piche-INAM	4.653					-	-	-	-	
Hargreaves-Samani (1985)	4.838	1.519	2.024	-2.994	1.580	0.619	0.618	0.383	Bad	High
P-M: Oliveira and Carvalho										
(1998)	3.925	1.567	2.196	-2.928	1.873	0.619	0.545	0.338	Bad	High
Budyko (1950)	4.830	1.699	2.224	-10.90	3.220	0.604	0.397	0.240	Bad	High
Holdrige	3.888	1.707	2.386	-10.90	4.000	0.604	0.397	0.240	Bad	High
Blaney-Morin	2.002	2.719	3.337	-1.181	2.914	0.755	0.507	0.383	Bad	Very high
Modified Hargreaves	3.300	1.718	2.352	-1.095	1.741	0.691	0.609	0.421	Bad	High
Ivanov (1977)	3.276	1.543	2.089	1.139	1.072	0.773	0.778	0.601	Sufferable	Very high
McGuiness-Bordne (2005)	7.015	2.792	3.250	-6.568	1.599	0.457	0.510	0.233	Bad	Moderate
Camargo (1971)	3.952	1.724	2.386	-5.750	2.632	0.485	0.375	0.182	Bad	Moderate
Hamon (1961)	3.918	1.570	2.109	-2.227	1.755	0.662	0.622	0.412	Bad	High
Kharrufa (1985)	6.166	2.181	2.619	-5.954	1.720	0.548	0.557	0.305	Bad	High
Benevides-Lopez (1970)	4.145	1.340	1.804	-1.647	1.520	0.758	0.753	0.571	Sufferable	Very high
McCloud (2001)	5.006	1.486	1.868	-0.920	1.113	0.672	0.758	0.509	Bad	High
Romanenko mod Oudth et										
al (2005)	3.931	1.222	1.739	1.139	0.893	0.773	0.852	0.658	Average	Very high
Schendel (1967)	5.070	1.414	1.780	-2.208	1.353	0.738	0.765	0.565	Sufferable	Very high

FAO Blaney-Criddle										
(1950)	2.664	2.354	3.165	-0.584	1.965	0.059	0.423	0.025	Bad	Very low
Linacre (1977)	3.634	1.524	2.055	-1.325	1.645	0.750	0.697	0.523	Sufferable	Very high
Trajkovic (2007)	6.197	2.372	2.896	1.180	0.561	0.178	0.451	0.081	Bad	Low
Ravazzani et al (2012)	7.795	3.482	3.977	0.009	0.596	0.201	0.448	0.090	Bad	Low
Dalton (1802)	6.421	2.413	3.365	1.820	0.441	0.759	0.751	0.570	Sufferable	Very high
Trabert (1896)	5.848	1.977	2.754	1.745	0.497	0.771	0.802	0.618	Average	Very high
Meyer (1926)	5.290	2.019	2.738	2.233	0.457	0.640	0.764	0.489	bad	High
Rohwer (1931)	7.088	2.971	4.132	1.899	0.389	0.752	0.694	0.522	Sufferable	Very high
Penman (1948)	1.823	2.846	3.329	1.412	1.778	0.782	0.565	0.441	Bad	Very high
Albrecht (1950)	16.685	12.193	15.990	2.226	0.146	0.717	0.299	0.214	Bad	Very high
Brokamp et al. (1963)	9.437	5.131	6.553	1.703	0.313	0.774	0.555	0.429	Bad	Very high
WMO (1966)	5.929	2.227	3.181	2.097	0.431	0.732	0.764	0.559	Sufferable	Very high
Mahringer (1970)	5.438	1.733	2.412	1.745	0.535	0.771	0.830	0.640	Average	Very high
H-Modif by Bert et. All										
(2014)	4.223	1.502	2.090	-2.838	1.774	0.621	0.576	0.358	Bad	High

(1) *Linear coefficient differs from zero by t-test at 5% probability level.

(2) *Angular coefficient differs from one by t-test at 5% probability level.

(3) Performance of the CS coefficient (CAMARGO and SENTELHAS, 1997).

(4) Classification of person's linear correlation coefficient.

The average daily ETo ranged from 1.823 to 16.685 (mm.d^{-1}), with the Albretck (1950) method achieving the highest value and Penman (1948) the lowest (Table 3). Muhammad et al. (2019) in a study carried out in Peninsula Malaysia· observed an identical result with the Penman (1948) method, with an average of 0.44. Filho et al. (2010) states that an increase in wind speed results in an increase in ETo, explaining the effect of varying the wind speed at two meters on the ETo value. In turn, Allen et al. (1998) state that in humid climate regions the high humidity of the air and the presence of clouds make the ETo rate lower, a condition in which the wind replaces the saturated humid air with unsaturated air and, consequently, removes energy in the form of latent heat.

Although the Albrecht (1950) method achieved a very high correlation with a correlation coefficient of r=0.717, it performed very badly with a CS index of 0.214, an average ETo of 16.68 mm.day^{-1} and an average absolute error of 12.193 (Table 3). Therefore, this method proved to be unsuitable for estimating ETo in the conditions of the region under study, because it overestimates the ETo obtained from the INAM using the Piche evaporimeter method. A similar result was observed in a study carried out by Djaman *et al.* (2015) in Senegal, which found an average ETo value of 1.59, ranking eighth among the best ETo estimation methods.

With regard to the classification of the performance of the Budyko and Holdrige methods, which use only air temperature and are simpler, they performed very poorly when compared to the ETo-INAM with a CS performance coefficient of 0.24 for both respectively (Table 3). However, these methods had higher regression coefficients "a" and "b" than other methods, revealing some overestimation of ETo (Table 3). GURSKI *et al.* (2016) and Macuiza (2018) also obtained very poor performance for the Budyko method, with a CS index of 0.30 and 0.116 in the Curitiba-Brazil and Chimoio-Mozambique regions.

It was found that the methods of Hargreaves modified with index (CS=0.421), Hamon (1961) (CS=0.412), McCloud (2001) (CS= 0.509), Meyer (1926) (CS=0.489), Penman (CS=0.441) and Brokamp et al. (CS=0.429) performed poorly. The result verified by Macuiza (2018) contrasts with that verified in this study. For the methods of McCloud (2001) and Hamon (1961), he observed very poor performance for both, with CS indices of 0.202 and 0.098 respectively. On the other hand, the same author justifies the performance of the McCloud (2001) method because it only uses air temperature as an input variable.

The Ivanov (1977), Benevides Lopes (1970), Shendel (1967), Linacre (1977), Dalton (1802), Rohwer (1931) and WMO (1966) methods performed poorly with CS indices equal to (0.601), (0.571), (0.565), (0.523), (0.570), (0.522) and (0.559). The results found by Santana *et al.,* (2018) and (Júnior *et al., (*2012) were similar, having observed the method of Benevides-Lopez (CS=0.49) and Linacre (CS=0.49), a research carried out in the southern region of Maranhão over a period of 18 years.

The ETo estimated by the methods of Romanenko modified by Oudith (2005), Trabert (1896) and Mahringer (1970) performed better than the other methods discussed in this research, with CS indices of 0.658, 0.640 and 0.618 considered average. It should be noted that these methods had low MSE values and a "very high" correlation with "r" equal to 0.773, 0.771 and 0.771 (Table 3). Of the three methods, only Romanenko's method modified by Oudith (2005) does not use wind speed as an input variable. In a study carried out by Djaman *et al.* (2015) in Senegal, they found EAM values close to the results found in this research, with 2.78, 1.8 and 2.68 respectively.

Romanenko's method is an alternative for quantifying ETo when only meteorological data on air temperature and relative humidity is available from the local weather station. According to the findings of Queiroz *et al.* (2011), in studies carried out in the semi-arid region of Pernambuco and in the municipality of Barbalha-Brazil, respectively, they recommend the use of a method with a CS performance classification considered to be average and poor, as an alternative for estimating the need for irrigation.

The methods of Meyer (1926), Albrecht (1950), WMO (1966), Rohwer (1931), Dalton (1802), Mahringer (1970), Trabert (1896), Brocamp *et al,* (1963) and Penman (1948), which use wind speed as an input variable, showed a similar trend in data dispersion, with the coefficient (R^2) varying from 0.4091 to 0.6113 (Figure 4.21 to 4.28), with the former having the minimum value and the latter the maximum. A study by Muhammad et al. (2019) found that the WMO 1996 and Mahringer (1970) methods underestimated the ETo estimated by the standard method, ranking 8th and 11th among the methods studied.

4.3. Evaluation of ETo estimation methods using ETo estimated by the Hargreaves-Samani method (1985) as a standard.

The estimation of reference evapotranspiration by the methods of Piche-INAM, Penman-Monteith adjusted by Oliveira and Carvalho, Budyko, Holdrige, Blaney-Morin, modified Hargreave, Ivanov, McGuiness-Bordne, Camargo, Hamon, Kharufa, Benevides-Lopes, McCloud, Romanenko, Shendel, Blaney-Criddle, Linacre, Trajkovic, Ravazzani, Dalton, Trabert, Meyer, Rohver, Penman, Albrecht, Brokamp et al., WMO, Mahringer and Hargreaves de Bert et al. considering ETo estimated by the Hargreaves-Samani method as the standard is shown in Figure 5.

The daily ETo values, the errors (EAM and RQME), Pearson's correlation coefficient "r" and determination coefficient "R^2", linear coefficient "a" and angular coefficient "b", agreement index "d" and performance coefficient "CS" of each method in relation to ETo-INAM are shown in Table 4.

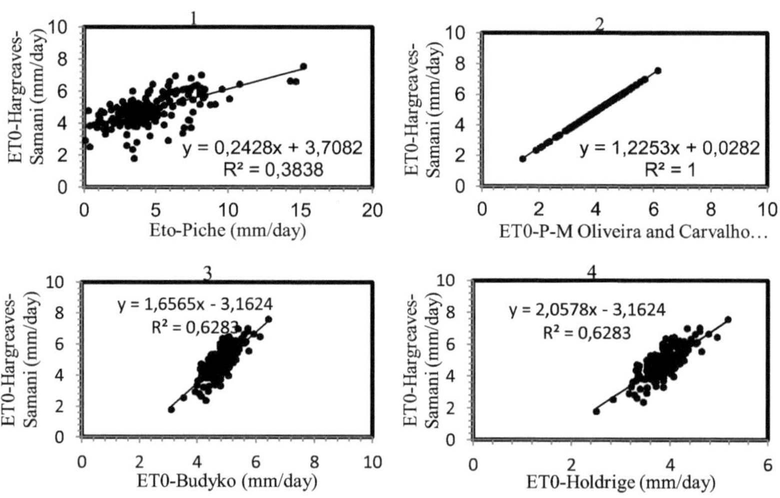

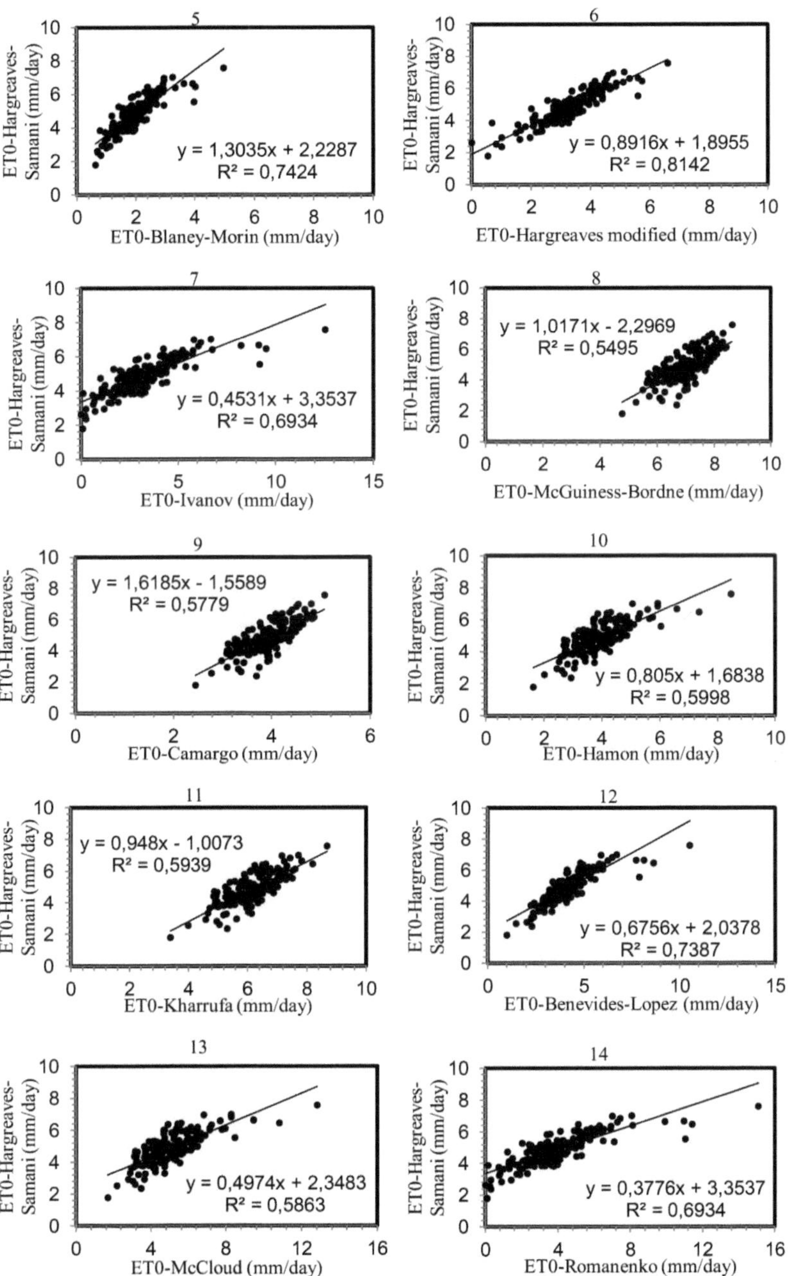

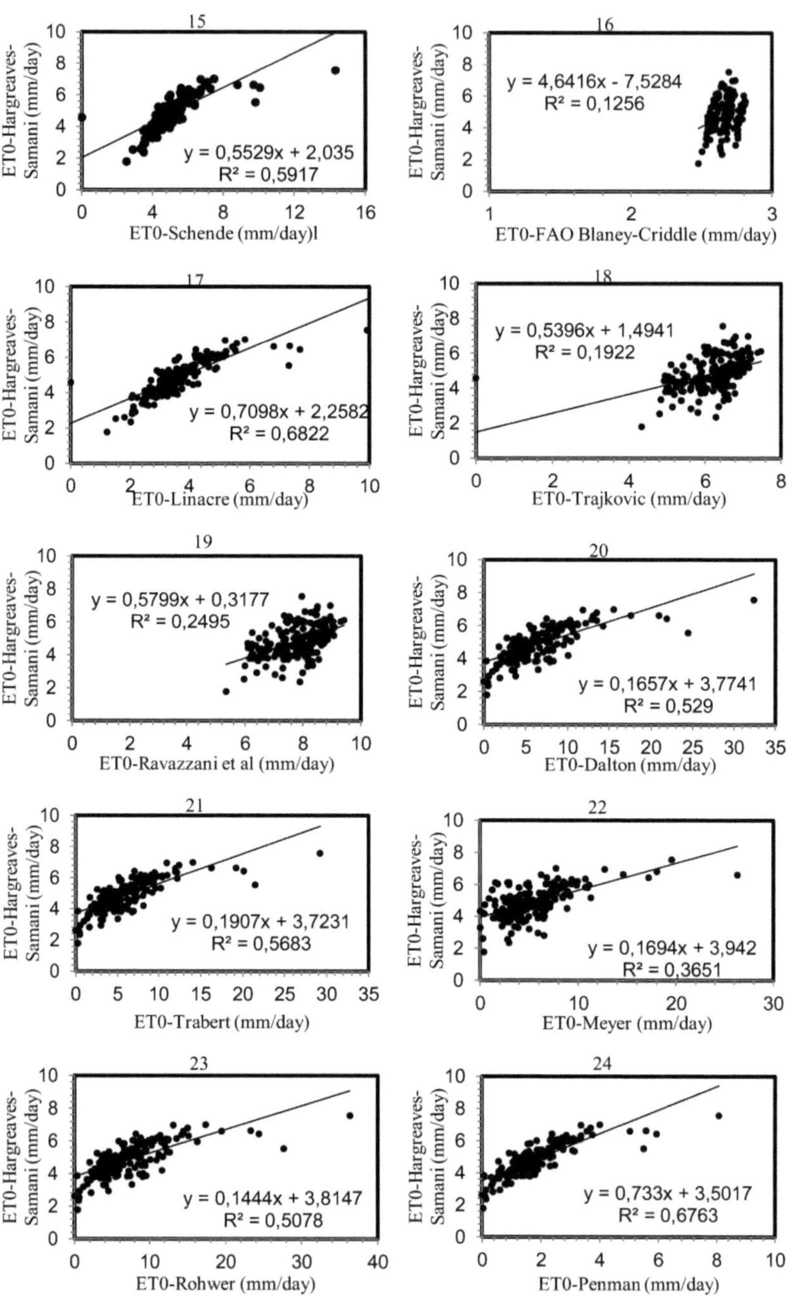

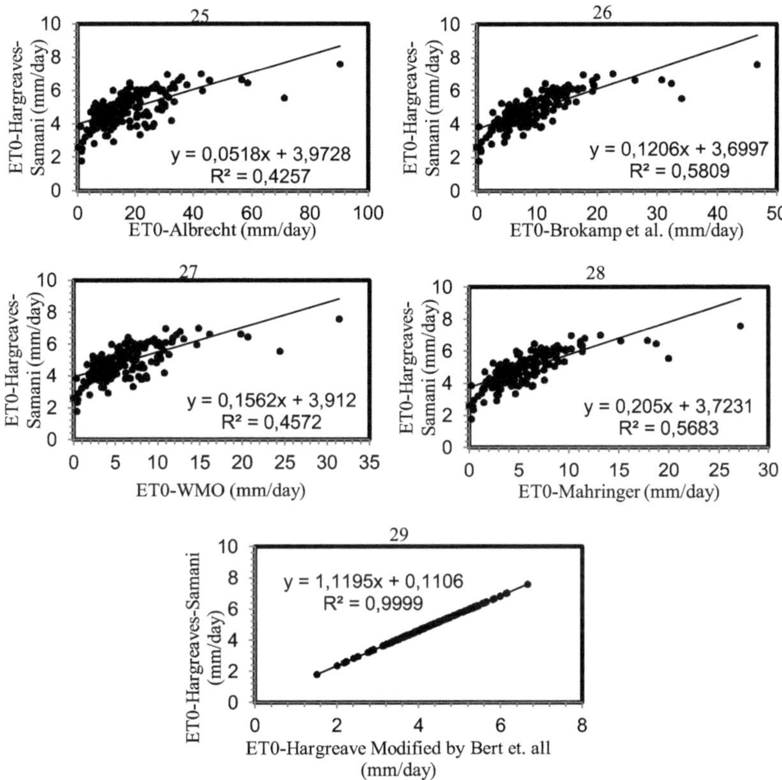

Figure 5: Linear regression of ETo values estimated by the Hargraves-Samani (1985) method, compared with ETo values estimated by the Piche, Penman Montheith adjusted by Oliveira and Carvalho (1998), Budyko, Holdrige, Blaney-Morin, modified Hargraves, Ivanov, McGuiness-Bordne, Camargo, Hamon, Kharrufa, Benevides-Lopez, McCloud, Romanenko, Shendel (1967), FAO-Blaney-Cride, Linacre (1977), Trajkovic (2007), Ravazzani et al. (2012), Dalton, Trabert (1896), Meyer, Rhwer, Penman, Albrecht, Brokamp et al, WMO, Mahringer (1970) and Hargreaves Modified by Bert et al (2014), for 183 days, from October 2019 to March 2020. Source: Author (2020).

Table 4: ETo (mm day^{-1}) of different methods, EAM, RQME, Pearson's linear coefficient (r), linear (a) and angular (b) coefficients obtained in the linear regression between ETo estimated by the Hargreaves-Samani (1985) method (mm day-1) (dependent variable) and by different methods (independent variable). Index (d) by Willmott *et al*. (1985) and CS index by Camargo and Sentelhas (1997) calculated based on estimated and observed ETo (mm), for the period from October 2019 to March 2020. Source: Author (2020).

Method	ET0	EAM	RQME	a[1]	b[2]	r	d	CS	Performance[3]	Correlation[4]
Hargreaves-Samani (1985)	4.838					-	-	-		
Eto-Piche	4.653	1.519	2.024	0.243	3.708	0.619	0.614	0.381	Bad	High
P-M: Oliveira and Carvalho (1998)	3.925	0.913	0.930	1.225	0.028	1.000	0.790	0.790	Very good	Almost perfect
Buddha (1950)	4.830	0.533	0.663	1.657	-3.162	0.793	0.769	0.610	Sufferable	Very high
Holdrige	3.888	1.016	1.185	2.058	-3.162	0.793	0.592	0.469	Bad	Very high
Blaney-Morin	2.002	2.836	2.885	1.304	2.229	0.862	0.389	0.335	Bad	Very high
Modified Hargreaves	3.300	0.008	0.111	0.892	1.896	0.902	0.704	0.635	Average	Almost perfect
Ivanov (1977)	3.276	1.746	1.916	0.453	3.354	0.833	0.633	0.527	Sufferable	Very high
McGuiness-Bordne (2005)	7.015	2.177	2.271	1.017	-2.297	0.741	0.457	0.339	Bad	Very high
Camargo (1971)	3.952	0.963	1.121	0.619	-1.559	0.760	0.617	0.469	Bad	Very high
Hamon (1961)	3.918	1.252	1.119	0.805	1.684	0.774	0.948	0.735	Good	Very high
Kharrufa (1985)	6.166	1.329	1.464	0.948	-1.007	0.771	0.611	0.471	Bad	Very high
Benevides-Lopez (1970)	4.145	0.819	0.940	0.676	2.038	0.860	0.829	0.712	Good	Very high
McCloud (2001)	5.006	0.728	0.987	0.497	2.348	0.766	0.823	0.630	Average	Very high
Romanenko mod Oudth et al (2005)	3.931	1.329	1.694	0.378	3.354	0.833	0.713	0.594	Sufferable	Very high

Schendel (1967)	5.070	0.458	0.892	0.553	2.035	0.769	0.835	0.643	Average	Very high
FAO Blaney-Criddle (1950)	2.664	2.185	2.369	4.642	-7.528	0.354	0.376	0.133	Bad	Moderate
Linacre (1977)	3.634	1.272	1.361	0.710	2.258	0.826	0.693	0.572	Sufferable	Very high
Trajkovic (2007)	6.197	1.438	1.653	0.540	1.494	0.438	0.505	0.221	Bad	Moderate
Ravazzani et al (2012)	7.795	2.957	3.093	0.580	0.318	0.499	0.354	0.177	Bad	Moderate
Dalton (1802)	6.421	2.429	3.934	0.166	3.774	0.727	0.449	0.327	Bad	Very high
Trabert (1896)	5.848	2.045	3.314	0.191	3.723	0.754	0.512	0.386	Bad	Very high
Meyer (1926)	5.290	1.984	2.999	0.169	3.942	0.604	0.499	0.302	Bad	High
Rohwer (1931)	7.088	2.932	4.709	0.144	3.815	0.713	0.390	0.278	Bad	Very high
Penman (1948)	1.823	3.021	3.078	0.733	3.502	0.822	0.400	0.329	Bad	Very high
Albrecht (1950)	16.685	11.96	16.547	0.052	3.973	0.652	0.125	0.081	Bad	High
Brokamp et al. (1963)	9.437	4.879	7.098	0.121	3.700	0.762	0.284	0.216	Bad	Very high
WMO (1966)	5.929	2.359	3.763	0.156	3.912	0.676	0.453	0.306	Bad	High
Mahringer (1970)	5.438	1.857	2.957	0.205	3.723	0.754	0.549	0.414	Bad	Very high
Harg-Modif by Bert et. All (2014)	4.223	0.615	0.624	1.120	0.111	1.000	0.897	0.897	Great	Almost perfect

(1) *Linear coefficient differs from zero by t-test at 5% probability level.

(2) *Angular coefficient differs from one by t-test at 5% probability level.

(3) Performance of the CS coefficient (CAMARGO and SENTELHAS, 1997).

(4) Classification of the person correlation coefficient.

Table 4 shows that the Albretch (1950) method was the one with the highest average daily ETo (16.685 mm/day) and the furthest from the standard method (4.838 mm/day). The same method also had the lowest agreement index (d=0.125) of the other methods, justifying its performance as very poor. Jerszurki (2016) points out that an increase in wind speed increases the evaporation rate of water, as the wind removes the water content in the air, since the wind speed recorded during the study period reached a maximum value of 20.6 m/s (Figure 4C).

The other methods that performed very poorly are: FAO Blaney Cridle (1950), Ravazzani et al. (2012), Brokamp et al. (1963), Trajkovic (2007), Rohwer (1931), Meyer (1926), WMO (1966), Dalton (1802), Penman (1948), Blaney-Morin, McGuiness Bordne (2005), Piche-INAM, and Trabert (1896) respectively. The McGuiness-Bordne method performed better in the study carried out in Mato Grosso, and was recommended by Tanaka *et al.* (2016) as an alternative to be used in irrigation management, a result that differs from that found in this research. The result obtained by Macuiza (2018) in the city of Chimoio, converges with that of this research, having obtained very poor performance for the McGuiness-Bordne (2005) and Blaney-Morin methods. Although the FAO24-Blaney-Criddle method is very practical, it proved to be unsuitable for estimating ETo on a daily scale. Júnior *et al.* (2011) evaluated the performance of this method in relation to Penman-Monteith and obtained a correlation coefficient greater than 0.8 in 15 of the 16 locations studied in Australia, for monthly ETo estimates.

It was observed that only the Benevides Lopes (1970) and Hamon (1961) methods performed well. The result found by Júnior *et al.,* (2012) in Aquidauana contradicts that observed in this research. In their study, the aim was to evaluate the performance of five empirical methods for estimating ETo on the daily, fortnightly and monthly scales, compared with the standard Penman-Monteith FAO-56 method. For the Hamon method, he found poor performance (daily scale), good performance (fortnightly scale) and very good performance (monthly scale). For Benevides Lopes, he found poor performance (daily scale) and average performance (fortnightly and monthly scale) for that region. In this study, Hamon's method proved to be the best alternative for estimating ETo due to its simplicity.

The Hargreave methods modified by Bert *et al.* (2014) and P-M adjusted by Oliveira and Carvalho (1988), underestimated ETo values estimated by the Hargreave

Samani method (1985) throughout the period studied (Figure 5.2 and 5.29), resulting in angular coefficients "a" close to one (1.1195 and 1.2253). The very good performance obtained by the P-M method adjusted by Oliveira and Carvalho is also justified by the coefficient of determination "R^2 " equal to one, as can be seen in Figure 5.2.

Analyzing the values in Table 3, it can be seen that the Hargreave method modified by Bert *et al.* (2014) showed the best performance index CS=0.897. The optimum performance of this method was expected as it is similar to the standard method and requires practically the same climatic elements; however, its application ends up being a better alternative for locations where air temperature and relative humidity are available. A similar result was observed in a study carried out in Senegal by Djaman, et al., (2015) where the angular coefficient value (1.69) was close to that found in this research.

4.4.   Evaluation of ETo estimation methods using the standard ETo estimated by the Penman-Monteith method adjusted by Oliveira and Carvalho (1998).

The estimation of reference evapotranspiration by the Piche-INAM, Hargreaves-Samani, Budyko, Holdrige, Blaney-Morin, modified Hargreave, Ivanov, McGuiness-Bordne, Camargo, Hamon, Kharufa, Benevides-Lopes, McCloud, Romanenko, Shendel, Blaney-Criddle, Linacre, Trajkovic, Ravazzani, Dalton, Trabert, Meyer, Rohver, Penman, Albrecht, Brokamp et al., WMO, Mahringer and Hargreaves de Bert et al. considering the ETo estimated by the Penman-Monteith method adjusted by Oliveira and Carvalho (1998) as the standard is shown in Figure 6.

The daily ETo values, the errors (EAM and RQME), Pearson's correlation coefficient "r" and determination coefficient "R2", linear coefficient "a" and angular coefficient "b", concordance index "d" and performance coefficient "CS" of each method in relation to the ETo- Penman-Monteith adjusted by Oliveira and Carvalho (1998) are shown in Table 5.

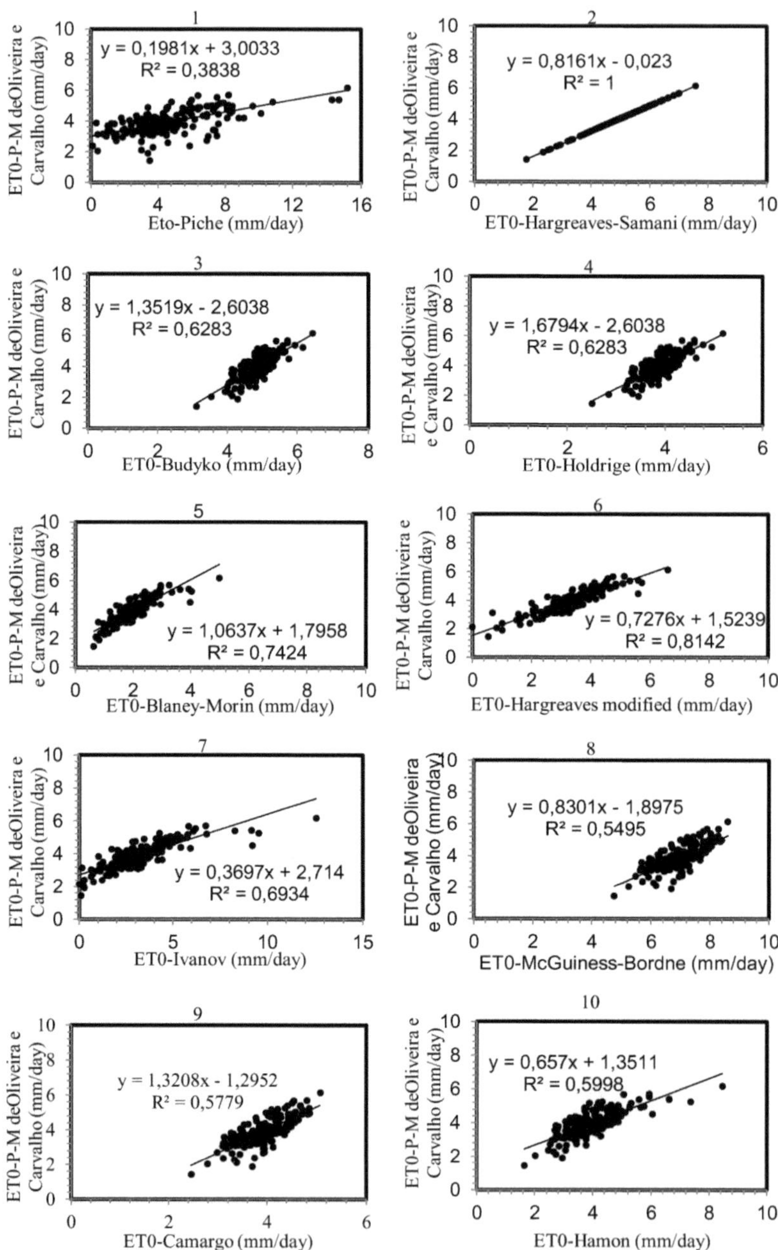

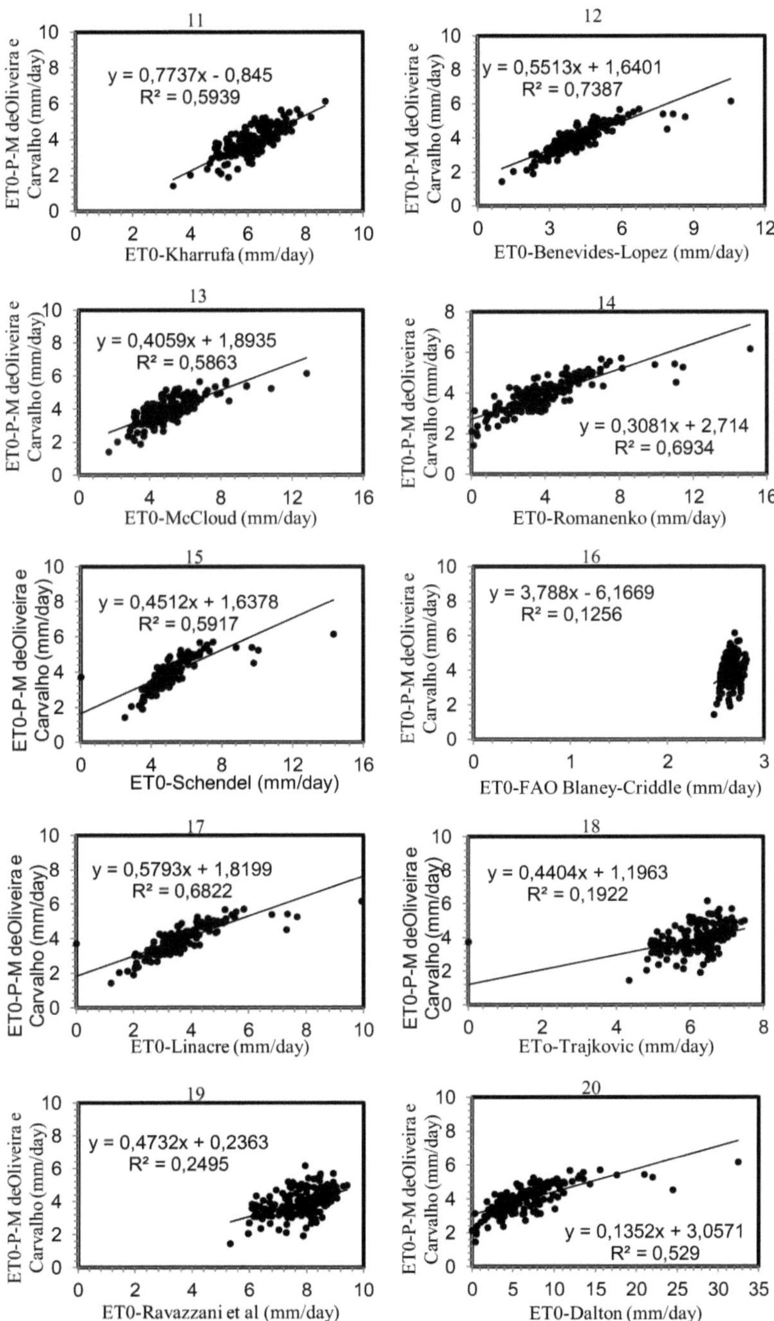

11
y = 0,7737x - 0,845
R² = 0,5939
ET0-P-M deOliveira e Carvalho (mm/day)
ET0-Kharrufa (mm/day)

12
y = 0,5513x + 1,6401
R² = 0,7387
ET0-P-M deOliveira e Carvalho (mm/day)
ET0-Benevides-Lopez (mm/day)

13
y = 0,4059x + 1,8935
R² = 0,5863
ET0-P-M deOliveira e Carvalho (mm/day)
ET0-McCloud (mm/day)

14
y = 0,3081x + 2,714
R² = 0,6934
ET0-P-M deOliveira e Carvalho (mm/day)
ET0-Romanenko (mm/day)

15
y = 0,4512x + 1,6378
R² = 0,5917
ET0-P-M deOliveira e Carvalho (mm/day)
ET0-Schendel (mm/day)

16
y = 3,788x - 6,1669
R² = 0,1256
ET0-P-M deOliveira e Carvalho (mm/day)
ET0-FAO Blaney-Criddle (mm/day)

17
y = 0,5793x + 1,8199
R² = 0,6822
ET0-P-M deOliveira e Carvalho (mm/day)
ET0-Linacre (mm/day)

18
y = 0,4404x + 1,1963
R² = 0,1922
ET0-P-M deOliveira e Carvalho (mm/day)
ETo-Trajkovic (mm/day)

19
y = 0,4732x + 0,2363
R² = 0,2495
ET0-P-M deOliveira e Carvalho (mm/day)
ET0-Ravazzani et al (mm/day)

20
y = 0,1352x + 3,0571
R² = 0,529
ET0-P-M deOliveira e Carvalho (mm/day)
ET0-Dalton (mm/day)

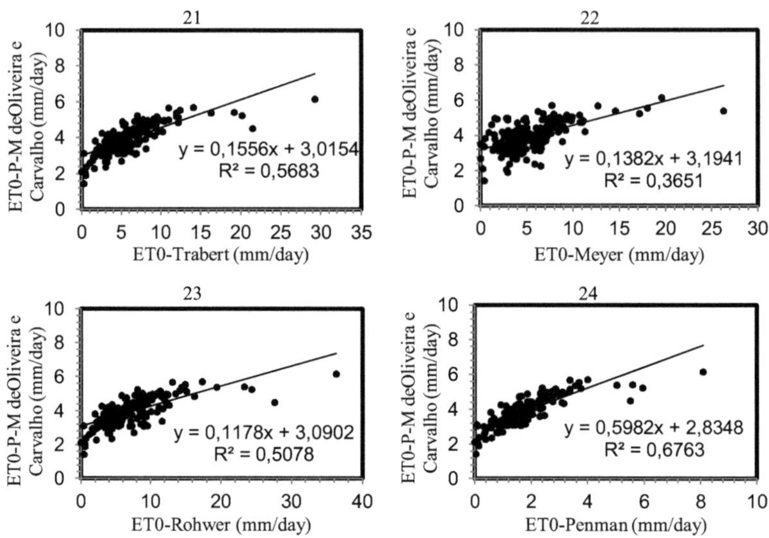

21
ET0-P-M deOliveira e Carvalho (mm/day)
ET0-Trabert (mm/day)
y = 0,1556x + 3,0154
R² = 0,5683

22
ET0-P-M deOliveira e Carvalho (mm/day)
ET0-Meyer (mm/day)
y = 0,1382x + 3,1941
R² = 0,3651

23
ET0-P-M deOliveira e Carvalho (mm/day)
ET0-Rohwer (mm/day)
y = 0,1178x + 3,0902
R² = 0,5078

24
ET0-P-M deOliveira e Carvalho (mm/day)
ET0-Penman (mm/day)
y = 0,5982x + 2,8348
R² = 0,6763

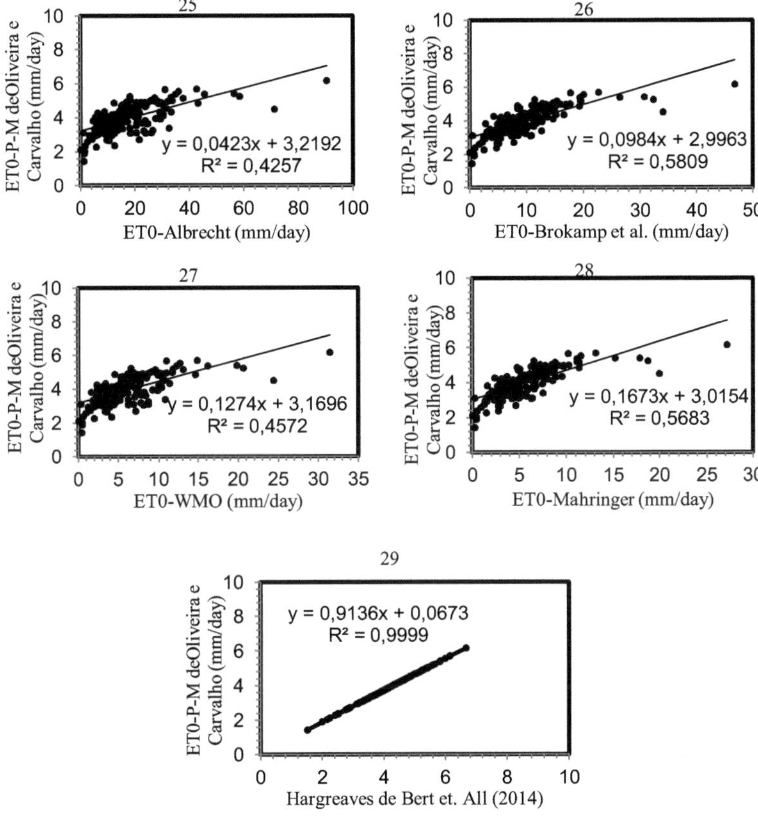

Figure 6: Linear regression of ETo values estimated by the Penman Montheith method adjusted by Oliveira and Carvalho (1998), compared with ETo values estimated by the Piche, Hargraves-Samani (1985), Budyko, Holdrige, Blaney-Morin, modified Hargraves, Ivanov, McGuiness-Bordne, Camargo, Hamon, Kharrufa, Benevides-Lopez, McCloud, Romanenko, Shendel (1967), FAO-Blaney-Cride, Linacre (1977), Trajkovic (2007), Ravazzani et al. (2012), Dalton, Trabert (1896), Meyer, Rhwer, Penman, Albrecht, Brokamp et al, WMO, Mahringer (1970) and Hargreaves Modified by Bert et al (2014), for 183 days, from October 2019 to March 2020. Source: Author (2020).

Table 5: ETo (mm day-) of different methods, EAM, RQME, Pearson's correlation coefficient (r) and coefficient of determination (R2), linear (a) and angular (b) coefficients obtained in the linear regression between ETo estimated by the Penman Monteith method adjusted by Oliveira and Carvalho (1998) (mm day-1) (dependent variable) and by different methods (independent variable). Index (d) by Willmott *et al.* (1985) and CS index by Camargo and Sentelhas (1997) calculated based on estimated and observed ETo (mm), for the period from October 2019 to March 2020. Source: Author (2020).

Models	ET0	EAM	RQME	a [1]	b [2]	r	d	CS	Performance [3]	Correlation [4]
P-M de Oliveira and Carvalho				-	-					-
(1998)	3.925			-	-	-	-	-	-	
Eto-Piche	4.653	1.567	2.196	0.198	3.003	0.619	0.530	0.328	Bad	High
Hargreaves-Samani (1985)	4.838	0.913	0.930	0.816	-0.023	1.000	0.788	0.788	Very good	Almost perfect
Budyko (1956)	4.830	0.917	1.037	1.352	-2.604	0.793	0.622	0.493	Bad	Very high
Holdrige	3.888	0.435	0.545	1.679	-2.604	0.793	0.765	0.606	Sufferable	Very high
Blaney-Morin	2.002	1.923	1.965	1.064	1.796	0.862	0.455	0.392	Bad	Very high
Modified Hargreaves	3.300	0.653	0.760	0.728	1.524	0.902	0.833	0.752	Good	Almost perfect
Ivanov (1977)	3.276	1.048	1.366	0.370	2.712	0.833	0.722	0.601	Sufferable	Very high
McGuiness-Bordne (2005)	7.015	3.089	3.137	0.830	-1.898	0.741	0.322	0.239	Bad	Very high
Camargo (1971)	3.952	0.419	0.533	1.321	-1.295	0.760	0.798	0.606	Sufferable	Very high
Hamon (1961)	3.918	0.463	0.592	0.657	1.351	0.774	0.870	0.674	Good	Very high
Kharrufa (1985)	6.166	2.240	2.303	0.774	-0.845	0.771	0.416	0.321	Bad	Very high
Benevides-Lopez (1970)	4.145	0.441	0.718	0.5513	1.640	0.860	0.868	0.746	Good	Very high
McCloud (2001)	5.006	1.135	1.485	0.4059	1.894	0.766	0.641	0.491	Bad	Very high
Romanenko (2005)	3.931	1.059	1.538	0.3081	2.714	0.833	0.707	0.588	Sufferable	Very high
Schendel (1967)	5.070	1.185	1.452	0.4512	1.638	0.769	0.633	0.487	Bad	Very high
FAO Blaney-Criddle (1950)	2.664	1.302	1.475	3.788	-6.167	0.354	0.427	0.151	Bad	Moderate

Linacre (1977)	3.634	0.493	0.711	0.5793	1.820	0.826	0.856	0.707	Good	Very high
Trajkovic (2007)	6.197	2.312	2.420	0.4404	1.196	0.438	0.366	0.160	Bad	Moderate
Ravazzani et al. (2012)	7.795	3.870	3.954	0.4732	0.236	0.499	0.257	0.128	Bad	Moderate
Dalton (1802)	6.421	2.860	4.470	0.1352	3.0571	0.727	0.347	0.253	Bad	Very high
Trabert (1896)	5.848	2.380	3.791	0.1556	3.0154	0.754	0.402	0.303	Bad	Very high
Meyer (1926)	5.290	2.166	3.329	0.1382	3.1941	0.604	0.401	0.242	Bad	High
Rohwer (1931)	7.088	3.450	5.293	0.1178	3.0902	0.713	0.299	0.213	Bad	Very high
Penman (1948)	1.823	2.144	2.193	0.5982	2.8348	0.822	0.464	0.381	Bad	Very high
Albrecht (1950)	16.685	12.826	17.285	0.0423	3.2192	0.652	0.099	0.064	Bad	High
Brokamp et al. (1963)	9.437	5.667	7.807	0.0984	2.9963	0.762	0.218	0.166	Bad	Very high
WMO (1966)	5.929	2.628	4.204	0.1274	3.1696	0.676	0.357	0.241	Bad	High
Mahringer (1970)	5.438	2.079	3.363	0.1673	3.0154	0.754	0.440	0.332	Bad	Very high
				0.9136	0.0673					Almost
Hargr. mod Bert et. All (2014)	4.223	2.380	3.791			1.000	0.967	0.967	Great	perfect

(1) *Linear coefficient differs from zero by t-test at 5% probability level.

(2) *Angular coefficient differs from one by t-test at 5% probability level.

(3) Performance of the CS coefficient (CAMARGO and SENTELHAS, 1997).

(4) Classification of the person correlation coefficient.

According to the CS performance classification of Camargo and Sentelhas (1997), the methods of Blaney-Morin, Penman (1948), Mahringer (1970), Piche-INAM, Kharrufa (1985), Trabert (1896), Dalton (1802), Meyer (1926), WMO (1966), McGuiness Bordne (2005), Rohwer (1931), Brokamp et al. (1963), Trajkovic (2007), FAO Blaney Cridle (1950), Ravazzani et al. (2012) and Albrecht (1950) were the worst performers (Table 5). The result found by Souza (2009) corroborates that found in this research for the Kharrufa (1985) method, which obtained a CS index of 0.15 for the daily ETo scale and was given a very poor performance in a study that aimed to compare eight ETo estimation equations for the municipality of Rio Branco-Acri. According to Sales (2008), the FAO Blaney-Cridle (1950) method showed higher values for standard deviation and coefficient of variation (0.99 and 0.96) compared to the other methods. The poor performance of the FAO Blaney-Cridle (1950) and Trajkovic (2007) methods observed in this research is also justified by the lower values of the coefficient of determination R^2 (0.1256 and 0.1922), as can be seen from the dispersion of the data in the graphs in figure (6.16 and 6.18). Also, Matos and Silva (2016) do not recommend using the FAO Blaney Cridle (1950) method because they found a low coefficient of determination (R^2 =0.57) in the municipality of Barbalha. Albrecht's method (1950) was the one with the highest MAS and RQME values (12.826 and 17.285), justifying its performance in this study. However, it is not recommended to use the methods that performed so poorly for estimating daily ETo.

The Budyko (1956), McCloud (2001) and Shendel (1967) methods performed poorly, achieving CS indices of 0.493, 0.491 and 0.487. Macuza (2018) observed a different result, obtaining poor and average performance for the Budyko and McCloud methods with CS indices of 0.533 and 0.618 respectively. The result found by Cunha et al (2013) differs from that found in this research. He obtained a CS index of 0.69 and classified it as good.

The methods of Romanenko (2005), Ivanov (1977) and Holdrige and Camargo (1971) performed poorly, obtaining CS indices of 0.588, 0.601 and 0.606 for the last two methods. Similar results were observed by Santana *et al.* (2018), Macuiza (2018), and Júnior *et al.* (2012), who in different studies performed "poorly" and Júnior *et al.* (2011), with the aim of evaluating methods for estimating reference evapotranspiration (ETo) in conditions of low and high relative humidity in the semi-arid Northeast of Brazil, observed "very poor" performance for the Camargo (1971) method in conditions of low relative humidity and "average" for conditions of high relative humidity, a result that differs from that found in this research. Lopes *et al.* (2018) also observed poor performance for the Ivanov (1977), Romanenko (2005) and Holdrige methods. Santana *et al* (2018) found a different result for the Ivanov method, with a CS index of 0.33 classified as "poor".

Table 5 shows that the Hamon (1961) (CS=0.674), Linacre (1977) (CS=0.707), Benevides Lopes (1970) (CS=0.746) and modified Hargreave (CS=0.752) methods performed well. In addition to their good performance, they were the ones with the very high to almost perfect correlation coefficient. This result was found by Júnior *et al.* (2011) in the conditions of the semi-arid Northeast of Brazil during a period of low relative humidity, having observed the Benevides Lopes (1970) method's performance as Sufferable. In the same study, Júnior *et al.* (2011) obtained good performance for the Benevides Lopes (1971) method during the period of high relative humidity, corroborating the result found in this research. Sousa *et al.* (2009), Matos, Silva (2016) and Santana *et al.,* (2018) performed poorly for the first and very poorly for the last two authors, for the Linacre (1977) methods. The result found in this research corroborates that observed by Lopes *et al.* (2018), for the Hamon method (1961), having achieved good performance with a CS index equal to 0.68 in the city of Areia PB in Brazil. In this way, the Hamon (1961) method, due to its practicality, becomes the alternative method for estimating ETo when there are difficulties in obtaining relative humidity, since this method only uses average air temperature and the number of possible hours of sunshine, which varies according to the month and latitude.

The Hargreaves Samani (1985) method performed very well, with a CS index of 0.788, so this method slightly overestimated the ETo values compared to the P-M method adjusted by Oliveira and Carvalho (1998) (Table 5). In addition to the overestimate, this method was the only one to have coefficients "r" and "R^2 " equal to one. A similar result was obtained by Macuiza, (2018), in a study carried out in Mozambique, where the CS index was 0.783. However, the result observed by Junior et al. (2011) contradicts that observed in this study, with poor performance in the semi-arid region of the Northeast.

Optimum performance was seen in the Hargreave method modified by Bert et al. (2014), which had a correlation (r=0.967) close to one and was classified as almost perfect. For the Hargreave Samani method, the CS index was 0.788 and classified as very good. Júnio *et al.* (2010), found a similar result to this research, obtaining a CS of 0.77, with performance considered very good in the Aquidauana region. This method is the best alternative for estimating ETo when only meteorological data on air temperature and relative humidity is available from the local weather station.

V. CONCLUSIONS AND RECOMMENDATIONS

5.1. Conclusions

According to the results obtained in this work, it can be concluded that:

1. The empirical ETo estimation methods used in this study showed low precision and poor performance on a daily scale, considering ETo obtained by the Piche method as the standard.

2. The methods of Hagreave modified by Bert et al. (2014), Penman-Monteith adjusted by Oliveira and Carvalho (1998), Hamon (1961) and Benevides-Lopes (1970) showed good precision, correlation, optimal, very good and good performance on the daily scale, considering ETo estimated by the Hargreaves-Samani method (1985) as the standard.

3. The methods of Hagreave modified by Bert et al., (2014), Hargreaves-Samani (1985), Hargreaves modified, Hamon (1961), Benevides-Lopes (1970) and Linacre (1977) showed good precision and correlation, optimal, very good and good performance on the daily scale, considering ETo estimated by the Penman-Monteith method adjusted by Oliveira and Carvalho (1998) as the standard.

5.2. Recommendations

- For the conditions of the study, it is recommended to use the empirical methods of Hagreave modified by Bert et al., (2014), Hamon (1961), Benevides-Lopes, Penman-Monteith adjusted by Oliveira and Carvalho (1998), Hargreaves-Samani (1985), Linacre (1977) and modified Hargreaves for irrigation management.
- Researchers are advised to carry out further studies on estimating ETo based on empirical methods for irrigation management on a weekly, decadal, fortnightly and monthly scale, since some methods have been designed for larger scales.
- It is also recommended that studies be carried out to adjust empirical methods that use a smaller number of input variables to estimate ETo in the study region.

VI.   BIBLIOGRAPHICAL REFERENCES

1. ALENCAR, L. P.; SEDIYAMA, G. C.; MANTOVANI, E. C. Estimation of reference evapotranspiration (FAO standard ETo), for minas gerais, in the absence of some climatic data. **Revista Engenharia Agrícola, v.**35, n.1, p.39-50, 2015.

2. ALLEN, R. G.; PEREIRA, L. S.; RAES, D.; SMITH, M. **Crop evapotranspiration**: **guidelines for computing crop water requirements**. Rome: FAO, (FAO. Irrigation and drainage paper, 56), p.300, 1998.

3. ARAÚJO, W. F.; COSTA, S. A. A.; SANTOS, A. E. Comparison between methods for estimating reference evapotranspiration (ETo) for Boa Vista. **Revista Caatinga** (Mossoró, Brazil), v.20, n.4, p.84-88, 2007.

4. BACK, A. J. Performance of empirical methods based on air temperature for estimating reference evapotranspiration in Urussanga, sc. **Irriga, Botucatu,** v. 13, n. 4, p. 449-466, 2008.

5. BERNARDO, S. **Irrigation manual. 6.ed. Viçosa:** Federal University of Vicosa. 1995.

6. BERNARDO, S.; SOARES, A. A.; MANTOVANI, E. C. **Manual de irrigação**. 8th edition-Viçosa: Ed. UFV, P.625, 2013.

7. BORGES, A. C.; MENDIONDO, E. M. Comparison of empirical equations for estimating reference evapotranspiration in the Jacupiranga River Basin. **Revista Brasileira de Engenharia Agrícola e Ambiental,** v.11, n.3, p.293-300, 2007.

8. CAMARGO A. P. de; CAMARGO M. B. P. de. An Analytical Review of Potential Evapotranspiration. **Bragantia, Campinas**, v.59(2), p.125-137, 2000.

9. CAMARGO, A. P.; SENTELHAS, P.C. Evaluation of the performance of different methods for estimating potential evapotranspiration in the state of São Paulo, Brazil. **Revista Brasileira de Agrometeorologia**, Santa Maria, v.5, n.1, p.89-97, 1997.

10. CARLESSO, R.; PETRY, M. T.; ROSA, G. M da; HELDWEIN, A. B. **Uses and Benefits of Automatic Weather Data Collection in Agriculture**. Santa Maria: Ed. UFSM, 2007 165 P.

11. CARVALHO, D. F.; OLIVEIRA, L. F. C. **Planejamento e Manejo da água na Agricultura Irrigada**. 1st edition. Ed. UFV. 239 P, 2012.

12. CHIMENE, C. A; STUDART, T. M. De C; CAMPOS, J. N. Analysis of Evapotranspiration Estimation Methods using SEVAP software in Mozambique. In **XIII Northeast Water Resources Symposium,** 2016.

13. CONCEIÇÃO, M. A. F. Adjustment of the Hargreaves model to estimate reference evapotranspiration in northwestern São Paulo. **Revista Brasileira de Agricultura Irrigada.** v.7, nº. 5, 2013.

14. DJAMAN, K.; BALDEA, A. B.; SOW, A.; MULLERA B.; IRMAKB, S.; N'DIAYE, M. K.; MANNEH, B.; MOUKOUMBI, Y. D.; FUTAKUCHI, K.; SAITO, K. Evaluation of sixteen reference evapotranspiration methods under sahelian conditions in the Senegal River Valley. **Journal of Hydrology**: Regional Studies v.3 p.139-159. 2015

15. DOORENBOS, J.; PRUITT, W. O. **Guidelines for predicting crop water requirements.** Irrigation and Drainage Paper, 24 Rome: FAO, p.197, 1977.

16. FERNANDES, A. L. T.; FRAGA JÚNIOR, E. F.; TAKAY, B. Y. Evaluation of the Penman-Piche method for estimating reference evapotranspiration in Uberaba, MG. **Brazilian Journal of Agricultural and Environmental Engineering.** Campina Grande, PB. v.15, n.3, p.270-276. 2011.

17. FILHO, L. C. A.;, CARVALHO, L. G.; EVANGELISTA, A. W. P.; JÚNIOR, J. A. Spatial analysis of the influence of meteorological elements on reference evapotranspiration in minas gerais. **Revista Brasileira de Engenharia Agrícola e Ambiental,** v.14, n.12, p.1294 - 1303, 2010.

18. GONÇALVES, F. M.; FEITOSA, H. O.; CARVALHO, C. M.; GOMES FILHO, R. R.; VALNIR JÚNIOR, M. Comparison of methods for estimating reference evapotranspiration for the municipality of Sobral. **Revista Brasileira de Agricultura Irrigada**, Fortaleza, v. 3, n. 2, p. 71-77, 2009.

19. GURSKI, B.C.; SOUZA J.L.M.; JERSZURKI D.; BARROCA, M.V. Alternative methods for estimating reference evapotranspiration for the main climatic types in Paraná. **Convibra**, 2016.

20. JERSZURKI, D. **Water dynamics in the soil-plant-atmosphere continuum: topics in reference evapotranspiration and plant water availability.** 2016. Thesis (Postgraduate Degree in Soil Science, Area of Concentration Soil and Environment)- Federal University of Paraná, Curitiba, 2016.

21. JUNIOR, E. G. C; OLIVEIRA, A. D; ALMEIDA, B. M; SOBRINHO, J. E. Methods for estimating reference evapotranspiration for the conditions of the semi-arid Northeast. **Ciências Agrárias, Londrina**, v. 32, s.1, p. 1699-1708, 2011.

22. JÚNIOR, J. C. F. B.; ANJOS, R. J.; SILVA, T. J. A.; LIMA, J. R. S.; ANDRADE, C. L. T. Methods for estimating daily reference evapotranspiration for the micro-region of Garanhuns, PE. **Brazilian Journal of Agricultural and Environmental Engineering,** v.16, n.4, p.380-390, 2012.

23. LOGCHEM, B. V.; QUEFACE, A. J. Responding to Climate Change in Mozambique, Synthesis Report. **In Phase II of the INGC project**, Maputo, 2012.

24. LOPES, D. M.; FREITAS, N. S. S.; NACIMENTO, V. C.; SILVA, M. R.; BORGES, P. F. Estimation of Reference Evapotranspiration by Different Methods for the Municipality of Areia-Pb. **Scientific Technical Congress of Engineering and Agronomy-Contecc**, p.5, Maceió-AL, 2018.

25. LORENZI K. S. de.. **Reference evapotranspiration between the penman-monteith and thorthwaite methods in the state of santa catarina.** 2010. Monograph (Degree in Agronomy) - Federal University of Santa Catarina, Florianópolis SC, 2010.

26. MACUIZA, D. C. **Estimation of reference evapotranspiration based on empirical methods for irrigation management.** 2018. Monograph (Degree in Environmental Agricultural Engineering)-University of Zambezi, Chimoio City, 2018.

27. MANTOVANI, E. C.; BERNARDO, S.; PALARETTI, L. F. **Irrigation: Principles and Methods**. 3rd edition, Viçosa Editora UFV, 2013, 355p.

28. MATOS, R. M.; SILVA, P. F. Analysis of methods for estimating monthly reference evapotranspiration for the municipality of Barbalha-CE. **Agropecuária** Científica no Semiárido, v.12, n.1, p.10-21, 2016.

29. MEDEIROS, A. T. **Estimation of reference evapotranspiration from the Penman-Monteith equation, lysimeter measurements and empirical equations**, in Piraipaba, CE. Thesis (Doctorate in Irrigation and Drainage) - Luiz de Queiroz College of Agriculture. University of São Paulo, São Paulo, 2002.

30. MENDONÇA, J.C.; SOUSA, E.F.; BERNARDO, S.; DIAS, G.P.; GRIPPA, S. Comparison between methods for estimating reference evapotranspiration (ETo) in the Norte Fluminense region, RJ. **Revista Brasileira de Engenharia Agrícola e** Ambiental, Campina Grande, v. 7, n. 2, p. 275-279, 2003.

31. MOURA, A. R. C. **Estimation of reference evapotranspiration in an experimental basin in the Northeast region**. Dissertation (Master's degree in civil engineering) - Federal University of Pernambuco, Recife-Pernambuco, 2009.

32. MUHAMMAD, M. K. I.; NASHWAN, M. S.; SHAHID, S.; ISMAIL, T. B.; SONG, Y. H.; 3 CHUNG, E. S. Evaluation of Empirical Reference Evapotranspiration Models Using Compromise Programming: A Case Study of Peninsular Malaysia. **Sustainability** 2019, v.11, p.4267;

33. NETTO, A. O. A.; BASTOS, E. A. **Agronomic Principles of Irrigation.** 1st edition. Embrapa Brasília, DF. 2013.262 p.

34. OLIVEIRA, L. F. C.; CARVALHO, D. F.; ROMÃO, P. A.; CORTÊS, F. C. Comparative study of reference evapotranspiration estimation models for some locations in the state of Goiás and the Federal District. **Pesquisa Agropecuária Tropical**, 31 (2): 121-126, 2001.

35. OLIVEIRA, M. A. A.; CARVALHO, D. F. Estimation of reference evapotranspiration and supplementary irrigation demand for corn (*Zea mays* L.) in Seropédica and Campos, State of Rio de Janeiro. Campina Grande, PB. **Revista Brasileira de Engenharia Agrícola e Ambiental**, v.2, n.2, p.132-135, 1998.

36. PEREIRA, A. R.; ANGELOCCI, L. R.; SENTELHAS, P. C.; FOLEGATTI, M. V.; NOVA, N. A. V.; MAGGIOTTO, S. R.; PEREIRA, F. A. C. Substantiation of the day FAO-56 reference evapotranspiration with data from automatic and conventional weather stations. **Revista Brasileira de Agrometeorologia**, v.10, n.2, p.251, 2002.

37. PEREIRA, A. R.; VILLA NOVA, N. A.; SEDIYAMA, G. C. **Evapo(transpi)rtion**. Piracicaba: FEALQ, 183p, 1997

38. PERREIRA, A.R; PRUIT, W. O.; SANTOS, T. E.M. Adaptation of the Thornthwaite Scheme for estimating daily reference evapotranspiration. **Agricultural Water Management.** Amsterdam Netherlands, v.66, n.2, p.251-257, 2004.

39. PILAU, F. E G.; BATTISTI, R.; LUCINDO, S.; RIGHI, E. Z. Performance of reference evapotranspiration estimation methods in Frederico Westphalen and Palmeira das Missões, RS. **Ciência Rural.** Santa Maria. v.42, n.2,p.283-290. 2012

40. RIBEIRO, A. A.; SIMEÃO, M.; SANTOS, A. R. B. Comparison of methods for estimating reference evapotranspiration in the rainy and dry seasons in Piripiri (PI). **Revista Agrogeoambiental**, Pouso Alegre, v. 8, n. 3, p. 89-100, Sep. 2016.

41. RINCÓN A. C. **Estimating evapotranspiration using empirical methods: application to a tropical green roof.** 2018. Dissertation (Master's in Civil Engineering) - Federal University of Rio de Janeiro, Rio de Janeiro, 2018.

42. SANTANA, J. S.; LIMA, E. F.; SILVA, W. A.; FERNANDES, M. C.; RIBEIRO, M. I. D. Reference evapotranspiration (eto) estimation equations for the region of balsas-ma. **Enciclopédia biosfera**, Centro Científico Conhecer - Goiania, v.15 n.27; p1. 2018.

43. SANTOS, A. A. R. **Adjustment of Empirical Methods and Spatial Interpolation Models of Reference Evapotranspiration Applied to the Climatic Conditions of the State of Rio de Janeiro**. Dissertation (Master's Degree in Biosystems Engineering) - Universidade Federal Fluminense, Niterói - RJ, 2015.

44. SILVA, J.L.R; MONTENEGRO, A. A. A.; SANTOS, T. E.M.; SANTOS, E.S. Performance of different methods for estimating reference evapotranspiration for Fernando de Noronha. **Irriga**, Botucatu, v. 19, n. 3, p. 390-404, 2014.

45. SILVA, M. G.; OLIVEIRA, J. B.; LÊDO, E. R. F.; ARAÚJO, E. M.; ARAÚJO E. M. Estimation of ETo by the Penman-Monteith FAO 56 and Hargreaves-Samani methods from Tx and Tn data for SOBRAL and TAUÁ in CEARÁ. **Revista ACTA Tecnológica - Scientific Journal** - ISSN 1982-422X , Vol. 5, number 2, 2010.

46. SILVA, A. P. C.; SILVA, J. C.; SANTOS, R.; SANTOS, M. A. A.; SANTOS D. P.; SANTOS, M. A. L. Comparative analysis between two reference evapotranspiration (ETo) methods for the Agreste region of Alagoas. **XXV CONIRD - National Congress on Irrigation and Drainage**, November, 2015

47. SOUZA, M. L. A. de. **Comparison of methods for estimating reference evapotranspiration (ETo) in Rio Branco, Acre.** 2009. Dissertation (Master's Degree in Agronomy, Area of Concentration in Plant Production) - Federal University of Acre, Rio Branco, 2009.

48. STONE, L. F.; SILVEIRA, P. M. da. **Determining evapotranspiration for irrigation purposes.** Goiânia: EMBRAPA-CNPAF. 49p. Documentos, 55. 1995.

49. TANAKA, A. A.; SOUZA, A. P.; KLAR, A. E.; SILVA, A. C.; GOMES, A. W. A. Reference evapotranspiration estimated by simplified models for the State of Mato Grosso. **Pesquisa Agropecuária Tropical**, v.32, n.2, p.121-126, 2001.

50. TRAJKOVIC, S. Hargreaves versus Penman-Monteith under humid conditions. Journal of Irrigation and Drainage Engineering, v. 133, p.38-42, 2007.

51. VIAGEM, S. J. **Simulation of the Impact of Climate Change on Irrigated Agriculture in the Sussundenga-Mozambique Region.** (Master's thesis in Water

Resources and Sanitation) - Federal University of Rio Grande do Sul, Porto Alegre, Rio Grande do Sul. 2013.

52. WILLMOTT, C. J. **Some comments on evaluation of model performance, Bulletin of American Meteorological Society,** Boston, v.63, p. 1309-1313, 1982.

APPENDICES

Appendix 1Validation of the methods considering Piche ETo obtained by INAM as a standard.
Source: Author (2020)

Methods	P
Eto-Piche	*
Hargreaves-Samani (1985)	8.87091E-21*
Penman-Monteith adjusted by Oliveira and Carvalho (1998)	8.87091E-21*
Budyko (1950)	1.43173E-19*
Holdrige	1.43173E-19*
Blaney-Morin	5.08428E-35*
Modified Hargreaves	2.67265E-27*
Ivanov (1977)	1.51469E-37*
McGuiness-Bordne (2005)	7.92083E-11*
Camargo (1971)	3.65071E-12*
Hamon (1961)	1.89318E-24*
Kharrufa (1985)	9.65609E-16*
Benevides-Lopez (1970)	1.99943E-35*
McCloud (2001)	2.25035E-25*
Romanenko mod Oudth et al (2005)	1.51469E-37*
Schendel (1967)	4.87502E-23*
FAO Blaney-Criddle (1950)	9.51917E-33*
Linacre (1977)	0.428904112*
Trajkovic (2007)	2.40155E-34*
Ravazzani et al (2012)	0.015637853*
Dalton (1802)	0.006341385*
Trabert (1896)	1.35123E-35*
Meyer (1926)	2.91611E-37*
Rohwer (1931)	1.93279E-22*
Penman (1948)	1.45988E-34*
Albrecht (1950)	5.53142E-39*
Brokamp et al. (1963)	3.10084E-30*
WMO (1966)	9.03642E-38*
Mahringer (1970)	6.14441E-32*
Harg-Modif by Bert et. All (2014)	2.91611E-37*

Appendix 2Validation of the methods considering Hargreav Samani's ETo (1985) as the standard

Source: Author (2020)

Methods	P
Hargreaves-Samani (1985)	*
Eto-Piche	8.87091E-21*
P-M: Oliveira and Carvalho (1998)	0*
Buddha (1950)	9.45829E-41*
Holdrige	9.45829E-41*
Blaney-Morin	3.37552E-55*
Modified Hargreaves	4.51547E-68*
Ivanov (1977)	2.41959E-48*
McGuiness-Bordne (2005)	3.59529E-33*
Camargo (1971)	0.000189262*
Hamon (1961)	2.91275E-37*
Kharrufa (1985)	7.34545E-36*
Benevides-Lopez (1970)	1.57123E-36*
McCloud (2001)	1.57123E-36*
Romanenko mod Oudth et al (2005)	2.41959E-48*
Schendel (1967)	2.41959E-48*
FAO Blaney-Criddle (1950)	8.54128E-07*
Linacre (1977)	6.28389E-47*
Trajkovic (2007)	5.38211E-10*
Ravazzani et al (2012)	6.14091E-13*
Dalton (1802)	2.07139E-31*
Trabert (1896)	7.49366E-35*
Meyer (1926)	1.35008E-19*
Rohwer (1931)	1.14526E-29*
Penman (1948)	3.31671E-46*
Albrecht (1950)	1.42843E-23*
Brokamp et al. (1963)	5.11134E-36*
WMO (1966)	8.47642E-26*
Mahringer (1970)	7.49366E-35*
Harg-Modif by Bert et. All (2014)	0*

Appendix 3Validation of the methods considering ETo obtained by the P-M equation adjusted by Oliveira and Carvalho (1998) as standard Source: Author (2020)

Methods	P
Penman-Monteith adjusted by Oliveira and Carvalho (1998)	*
Eto-Piche	8.87091E-21*
Hargreaves-Samani (1985)	0*
Budyko (1950)	9.45829E-41*
Holdrige	9.45829E-41*
Blaney-Morin	9.45829E-41*
Modified Hargreaves	3.37552E-55*
Ivanov (1977)	4.51547E-68*
McGuiness-Bordne (2005)	2.41959E-48*
Camargo (1971)	3.59529E-33*
Hamon (1961)	9.75211E-36*
Kharrufa (1985)	2.91275E-37*
Benevides-Lopez (1970)	1.20666E-54*
McCloud (2001)	1.57123E-36*
Romanenko mod Oudth et al (2005)	2.41959E-48*
Schendel (1967)	4.74578E-37*
FAO Blaney-Criddle (1950)	8.54128E-07*
Linacre (1977)	6.28389E-47*
Trajkovic (2007)	5.38211E-10*
Ravazzani et al (2012)	6.14091E-13*
Dalton (1802)	2.07139E-31*
Trabert (1896)	7.49366E-35*
Meyer (1926)	1.35008E-19*
Rohwer (1931)	1.14526E-29*
Penman (1948)	3.31671E-46*
Albrecht (1950)	1.42843E-23*
Brokamp et al. (1963)	5.11134E-36*
WMO (1966)	8.47642E-26*
Mahringer (1970)	7.49366E-35*
Harg-Modif by Bert et. All (2014)	0*

ANNEX

INSTITUTO NACIONAL DE METEOROLOGIA

DELEGAÇÃO PROVINCIAL DE MANICA

TABELA RESUMO DE DADOS METEOROLÓGICOS DA CIDADE DE CHIMOIO

MÊS: OUTUBRO   ANO: 2019

Data	Temperatura (°C)			Humidade (%)	Evaporação (mm)	Veloc. Vento (m/s)
	Média	Máxima	Mínima	Média	Total	Média
1	20,8	27,4	17,8	82	4,9	7,6
2	15,6	17,1	14,8	99	3,5	20.6
3	17,7	19,2	15,1	98	0,4	10.3
4	22,0	28,3	17,1	81	7,6	3.0
5	22,4	30,0	16,0	72	3,4	5.7
6	22,9	30,1	16,2	57	6,7	6.6
7	23,9	31,8	16,3	65	7,3	4.6
8	26,1	34,9	18,8	69	7,5	6.8
9	26,6	35,5	20,0	66	7,2	7.8
10	25,2	35,6	19,2	72	8,4	7.8
11	19,7	22,3	17,4	90	5,9	15.3
12	20,2	24,6	17,2	86	1,4	9.7
13	22,2	28,0	16,2	74	1,9	4.3
14	21,0	26,6	15,0	71	3,9	6.6
15	20,9	26,7	15,5	76	4,6	7.2
16	20,6	26,3	15,8	74	3,9	9.0
17	20,7	26,9	14,8	70	4,4	9.0
18	21,9	28,7	15,9	70	4,8	7.0
19	24,0	31,5	17,3	70	4,3	4.9
20	25,8	33,4	18,9	70	5,7	6.6
21	28,0	36,1	19,3	64	6,0	7.0
22	26,4	32,8	21,7	74	8,1	5.4
23	26,6	34,7	20,1	68	5,5	10.8
24	28,6	37,8	20,4	64	7,5	7.0
25	30,8	38,0	23,8	49	10,8	7.1
26	29,7	38,1	22,3	49	14,3	7.3
27	29,7	37,7	22,0	54	14,7	6.5
28	32,2	40,9	22,4	36	15,2	8.1
29	28,8	34,4	23,0	47		9.4
30	24,0	28,9	20,8	78	7,4	13.8
31	24,6	30,3	21,1	77	4,0	8.8

MÊS: NOVEMBRO   ANO: 2019

Data	Temperatura (°C)			Humidade (%)	Evaporação (mm)	Veloc. Vento (m/s)
	Média	Máxima	Mínima	Média	Total	Média
1	22,9	28,0	19,4	80	5,0	12.1

Cidade de Chimoio, Rua do Aeroporto, Bairro Trangapasso, C.P 497, Tel: 20030431, Fax: 21029720 E-mail meteochimoio@tdm.com

Data	Média	Máxima	Mínima	Humidade (%) Média	Evaporação (mm) Total	Veloc. Vento (m/s) Média
3	24,3	32,4	16,4	64	5,9	6.8
4	25,1	33,6	18,0	55	7,9	6.3
5	22,8	29,2	17,8	70	9,1	9.7
6	23,9	31,7	17,4	66	6,1	7.8
7	27,0	35,6	18,6	64	6,4	6.6
8	28,6	37,5	21,5	61	8,2	7.7
9	25,2	31,5	21,4	77	8,8	11.4
10	24,2	29,5	20,2	76	4,8	11.2
11	25,4	32,9	19,5	72	4,5	7.0
12	24,8	31,5	19,4	71	8,3	9.6
13	25,0	32,7	18,4	69	6,9	9.0
14	24,9	33,7	20,5	80	6,4	6.8
15	23,3	32,1	20,5	80	3,5	8.4
16	24,8	29,6	20,3	81	2,7	7.3
17	25,0	30,1	21,0	81	3,1	12.8
18	26,8	32,7	22,3	78	3,3	5.6
19	26,4	31,6	22,0	84	4,5	8.4
20	23,8	31,6	20,4	88	3,8	4.8
21	24,2	30,7	20,0	86	1,8	5.7
22	25,2	32,7	21,8	81	2,7	3.9
23	26,2	31,0	22,2	80	3,3	3.7
24	24,3	31,0	21,8	86	4,2	5.0
25	24,9	30,7	21,0	81	2,8	4.4
26	23,8	29,1	20,6	82	4,7	6.4
27	23,0	28,0	19,2	83	3,5	6.2
28	24,6	30,6	17,2	67	4,5	5.4
29	25,0	31,4	17,4	66	8,1	5.6
30	24,0	30,8	16,2	64	6,4	5.6

MÊS: DEZEMBRO **ANO: 2019**

Data	Temperatura (°C) Média	Máxima	Mínima	Humidade (%) Média	Evaporação (mm) Total	Veloc. Vento (m/s) Média
1	26,6	33,0	20,0	68	8,2	4.4
2	26,8	33,2	21,8	73	8,1	8.8
3	27,5	34,2	21,1	65	7,7	8.1
4	27,4	33,4	21,1	65	9,6	6.0
5	20,6	24,2	17,9	95	7,5	15.0
6	20,5	23,8	17,5	91	0,9	11.8
7	23,9	31,1	17,2	76	3,4	4.7
8	24,9	32,0	19,6	81	5,7	6.9
9	24,4	29,5	20,3	84	5,3	7.4
10	23,9	28,5	21,9	88	5,0	8.5
11	21,5	22,7	20,3	98	3,4	7.4

Cidade de Chimoio, Rua do Aeroporto, Bairro Trangapasso, C P 497, Tel: 20030431; Fax: 21029720, E-mail meteochimoio@jdm.com

Data	Temperatura (°C)			Humidade (%)	Evaporação (mm)	Veloc. Vento (m/s)
	Média	Máxima	Mínima	Média	Total	Média
13	22,8	26,3	19,2	79	3,3	8.6
14	24,0	28,3	18,3	74	3,0	6.0
15	26,4	32,4	20,1	70	6,4	4.4
16	23,3	29,4	19,0	85	5,6	5.2
17	21,3	24,7	18,2	84	2,5	14.2
18	23,3	28,6	18,7	72	2,9	7.3
19	25,4	32,9	18,0	72	4,8	5.4
20	25,8	32,6	19,8	72	6,5	6.0
21	26,1	31,4	21,7	75	6,9	7.0
22	25,8	31,5	19,7	71	5,6	6.1
23	24,4	29,6	20,0	70	6,2	9.1
24	26,2	32,3	20,0	66	6.0	4.8
25	27,4	33,8	21,9	68	6,5	5.5
26	28,1	34,4	22,8	68	7,4	6.6
27	28,3	34,6	22,8	69	7,0	6.2
28	24,0	27,4	22,1	89	7,0	8.0
29	22,4	26,5	19,4	87	1,0	12.3
30	24,7	29,5	20,3	82	2,3	5.5
31	26,1	30,6	21,3	76	3,9	4.2

MÊS: **JANEIRO** ANO: **2020**

Data	Temperatura (°C)			Humidade (%)	Evaporação (mm)	Veloc. Vento (m/s)
	Média	Máxima	Mínima	Média	Total	Média
1	25,5	29,4	21,8		4,8	7.6
2	25,6	30,5	21,4	78	3,5	6.0
3	25,6	30,6	21,0	76	4,5	5.2
4	25,8	30,5	20,2	74	5,1	5.9
5	25,9	31,0	20,0	74	5,4	6.0
6	26,5	32,4	21,4	71	5,5	5.7
7	26,6	32,4	21,1	70	8,4	6.2
8	26,7	31,4	21,2	70	7,6	5.5
9	26,2	32,4	22,0	79	6,9	5.6
10	27,1	32,9	22,1	73	5,8	5.0
11	23,1	25,6	21,0	95	6,9	8.2
12	22,9	27,2	21,0	89	0,7	5.9
13	22,9	26,3	20,9	92	1,5	3.8
14	23,6	27,7	20,4	86	2,0	4.4
15	22,6	27,0	20,0	92	2,8	6.4
16	23,0	26,2	20,0	90	1,5	7.8
17	23,1	26,3	21,0	94	1,9	8.6
18	24,5	28,7	20,4	87	1,4	6.0
19	24,3	30,0	21,5	93	4,2	5.8
20	21,9	23,7	20,0	98	0,1	9.3

Cidade de Chimoio, Rua do Aeroporto, Bairro Trangapasso, C.P 497, Tel. 20030431, Fax: 21029720, E-mail meteochimoio@tdm.com

Data	Média	Máxima	Mínima	Humidade (%) Média	Evaporação (mm) Total	Veloc. Vento (m/s) Média
22	22,1	26,6	18,0	75	3,7	10.3
23	23,2	28,4	17,8	78	4,2	5.0
24	23,4	27,9	19,8	81	3,5	6.0
25	23,4	28,1	19,4	81	3,2	7.4
26	24,8	30,4	19,3	76	3,4	5.7
27	26,2	32,3	19,5	74	4,6	5.2
28	22,7	26,5	20,1	92	4,5	7.5
29	23,7	28,1	20,0	84	1.0	8.7
30	23,3	28,4	19,7	82	3,5	4.6
31	22,8	26,5	19,2	81	3,4	7.4

MÊS: FEVEREIRO   ANO: 2020

Data	Temperatura (°C)			Humidade (%)	Evaporação (mm)	Veloc. Vento (m/s)
	Média	Máxima	Mínima	Média	Total	Média
1	23,2	28,0	18,9	82	2,9	5.5
2	23,7	29,1	18,9	82	3,3	6.0
3	23,7	28,7	21,0	87	3,6	4.9
4	23,6	27,7	20,9	91	2,1	5.4
5	25,2	30,0	21,0	83	1,2	4.0
6	25,8	31,6	22,8	82	3,6	5.5
7	24,8	29,9	21,1	85	2,4	4.7
8	24,3	29,4	21,9	86	3,5	5.8
9	24,0	30,6	21,0	84	3,0	5.0
10	24,9	29,5	19,9	86	3,1	6.9
11	24,0	26,9	22,3	90	3,0	9.0
12	20,7	23,0	19,6	100	3,0	10.3
13	22,3	26,2	19,4	99	0,4	10.1
14	23,4	28,3	20,2	91	1,1	6.6
15	24,2	28,5	21,2	88	1,9	5.7
16	25,0	29,0	22,1	84	2,3	6.2
17	24,5	28,8	20,3	83	3,6	4.3
18	25,3	30,2	21,0	78	3,6	3.1
19	25,1	29,3	20,9	81	5,6	5.6
20	25,2	30,2	20,1	81	3,1	3.2
21	26,3	31,2	21,9	80	3,9	2.9
22	26,1	30,7	22,0	82	4,7	2.8
23	26,6	32,4	23,1	81	3,7	2.1
24	20,4	22,7	18,7	94	3,7	11.2
25	20,9	24,7	17,4	76	2,5	12.2
26	22,0	26,7	17,0	80	2,4	6.0
27	22,7	27,8	17,1	78	3,7	3.7
28	23,9	29,6	18,5	78	4,6	2.4
29	21,1	23,8	18,8	95	4,6	5.6

MÊS: **MARÇO**					ANO: **2020**
	Temperatura (°C)			Humidade (%)	Evaporação (mm)	Veloc. Vento (m/s)
Data	Média	Máxima	Mínima	Média	Total	Média
1	20,0	24,0	17,0	84	1,5	16.2
2	21,0	26,4	16,7	83	3,2	7.6
3	20,7	26,6	17,9	82	3,7	4.2
4	20,8	25,6	16,0	83	3,5	6.3
5	21,7	26,4	18,0	79	3,1	6.3
6	22,1	26,9	17,7	80	3,9	5.8
7	21,7	26,2	17,9	76	4,5	10.7
8	22,5	27,5	17,3	82	4,1	8.3
9	23,6	28,3	19,8	82	3,9	7.1
10	23,8	28,8	19,2	81	4,4	6.0
11	24,6	30,0	19,7	81	4,6	4.1
12	24,9	30,1	20,8	81	4,5	3.8
13	24,6	30,2	19,0	82	4,3	5.4
14	25,5	31,0	21,0	76	5,2	4.3
15	25,6	32,2	20,5	73	5,2	3.0
16	24,0	29,1	213	77	5,6	4.8
17	25,3	26,6	19,0	83	4,2	11.8
18	21,5	25,4	16,9	80	3,5	9.9
19	22,8	27,9	17,1	81	3,1	2.9
20	22,4	26,3	18,2	86	3,6	3.7
21	22,6	27,7	17,9	82	2,4	3.5
22	22,2	27,1	17,2	80	3,6	3.5
23	23,3	28,4	18,4	78	3,2	2.9
24	24,9	29,7	20,0	72	4,0	2.2
25	25,4	30,7	21,0	71	6,1	2.2
26	26,5	33,0	19,6	66	6,0	3.2
27	24,8	29,1	20,3	77	7,3	4.2
28	22	25,8	19,7	88	3,0	9.2
29	20,9	28,0	16,3	78	5,1	5.7
30	22,4	27,7	17,1	78	3,5	3.2
31	21,9	27,5	16,8	80	4,0	3.4

Chimoio, aos 24 de Abril de 2020

Cidade de Chimoio, Rua do Aeroporto, Bairro Trangapasso, C.P 497, Tel: 20030431, Fax: 21029720, E-mail meteochimoio@idm.com

Printed by Books on Demand GmbH, Norderstedt / Germany